Louisville

The Greatest City

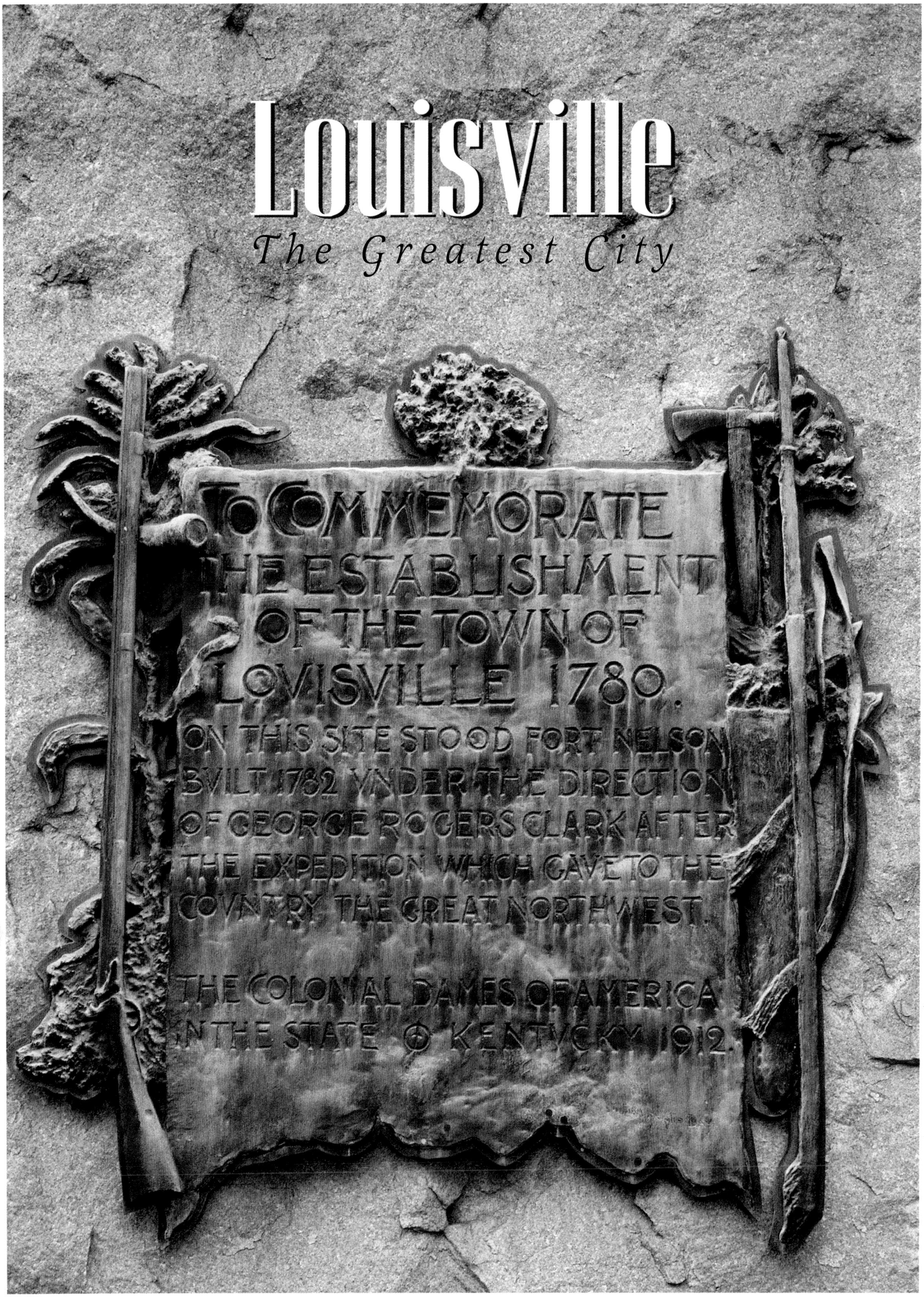

▲ © STEVE BAKER / HIGHLIGHT PHOTOGRAPHY

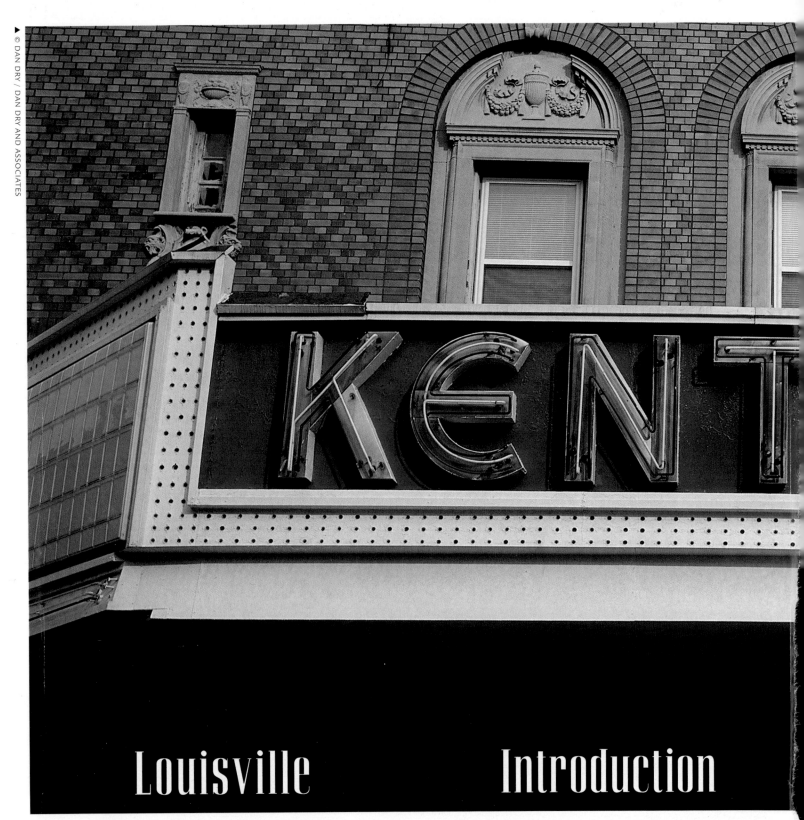

Louisville

The Greatest City

Introduction

Muhammad Ali

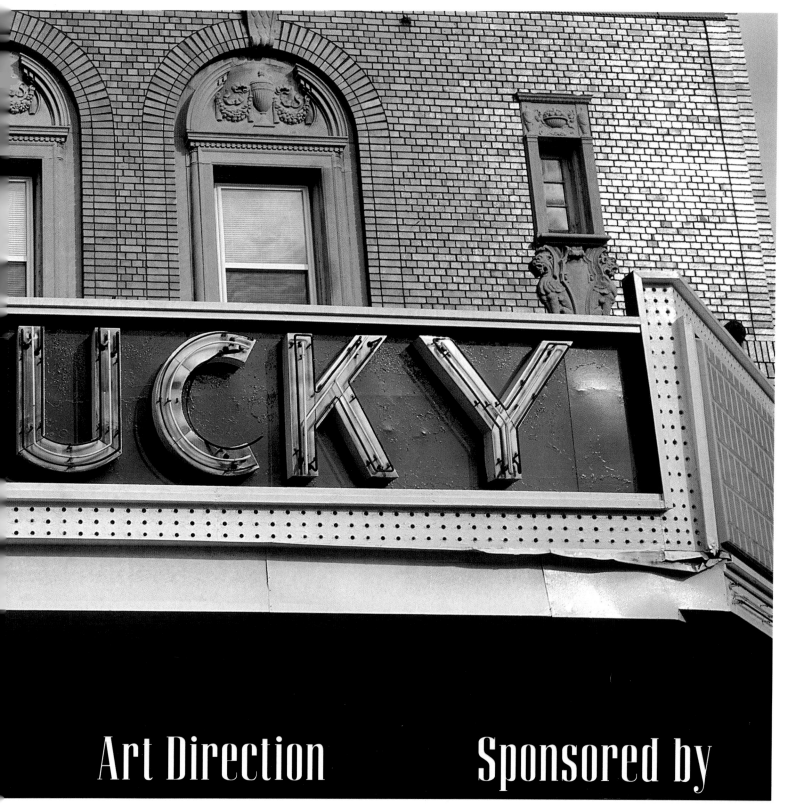

Art Direction

Bob Kimball

Sponsored by

Greater Louisville Inc.

© DAN DRY / DAN DRY AND ASSOCIATES

Louisville
The Greatest City

Introduction .... *Page 7*

Muhammad Ali captures the essence of Louisville as he reflects on the qualities that make his city great.

Photo-Essay .... *Page 12*

An enduring portrait of Louisville featuring images from the area's finest photographers.

Profiles in Excellence .... *Page 280*

A look at the corporations, businesses, professional groups, and community service organizations that have made this book possible.

Photographers .... *Page 428*

Index of Profiles .... *Page 430*

URBAN TAPESTRY SERIES
TOWERY PUBLISHING, INC.

Louisville is a city to love. Who could resist the charm of our town on the river? From the renowned Actors Theatre to the thriving downtown, attractions like Churchill Downs, Papa John's Cardinal Stadium, and the Louisville Zoo are all-time favorites.

Louisville is a city that is inextricably tied to the Ohio River, beginning with its original settlement at the Falls of the Ohio. Right now, the city is in the midst of an ambitious riverfront renaissance, with the recent development of its magnificent Waterfront Park, a seven-mile RiverWalk, and the newly opened Louisville Slugger Field—home of the RiverBats Triple-A baseball team.

© STEVE BAKER / HIGHLIGHT PHOTOGRAPHY

Louisville is also home to some of the most recognized icons of America, including the Kentucky Derby, the Louisville Slugger, and the city's most famous native son—me! The newly proposed Muhammad Ali Center will complement the Kentucky Derby Museum and the Louisville Slugger Factory and Museum in providing both residents and guests with state-of-the-art facilities highlighting what is uniquely Louisville.

The true strength of Louisville, however, can be found in the richness and diversity of its neighborhoods. Whether one is in the cast-iron historic district along West Main Street; the Victorian Old Louisville district; the Chickasaw, Shawnee, and Iroquois neighborhoods that surround Louisville's wonderful Olmsted parks; or the boulevards that connect these neighborhoods, the sense of community and family is always present. Louisville has always been able to maintain a nice combination of southern hospitality and midwestern work ethic. Such a combination is appreciated by both residents and visitors.

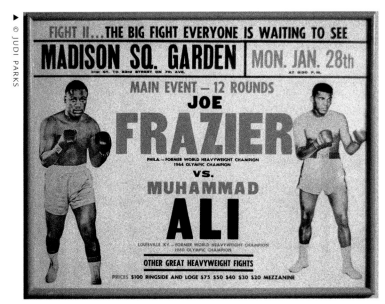

The living room of my boyhood home was where I first listened to boxing fights on the radio between great pugilists like Rocky Marciano and Joe Wolcott. Of course, it was also in Louisville that I got my start in boxing.

I have fond memories of growing up in the city. One of my favorite pastimes was hanging out at Albert's grocery store on 34th and Grand, every Saturday and Sunday night. They used to have quartets and singing groups on the corner there . . . it was so much fun. I could never have dreamed that there would be a street named in my honor—Muhammad Ali Boulevard—just minutes from where I played as a boy.

It is no surprise, then, that I have chosen to return to my hometown to build the Muhammad Ali Center. The vision of the Center is even greater than a museum about boxing or fame. It will appeal to the heart, spirit, and imagination of people. At the Center, we will actively engage children and adults in learning to make commitments to personal growth, discipline, tolerance, and respect for peace and love.

I am proud to be working in partnership with the University of Louisville to develop the Muhammad Ali Institute as part of the Center, focused on peacemaking, conflict resolution, and dispute mediation. Our partnership will foster the development of research and practical applications in these universally critical arenas.

My mission is to assemble scholars, activists, educators, and individuals from across the globe to advance their work in areas such as human rights, physical and spiritual wellness, and alleviation of world hunger. It is my hope that this Center will add to the unique flavor of Louisville, and that it will inspire children and adults everywhere to be as great as they can be.

Louisville is a special city, full of wonders and treasures. Never before has there been a book that reflected the city's greatness as majestically as this volume. If you're a native Kentuckian, relish the beautiful photos of places you know so well. If you're a visitor, welcome! Relax in our southern hospitality and comfort. As you will soon see, visitors aren't strangers here for long. Read this book and fall in love with Louisville.

In peace,
Muhammad Ali

Despite national concerns over health hazards, the tobacco industry thrives in Kentucky, the nation's second-largest producer of the crop. The commodity has a storied history in the state, where farms and drying barns dot the landscape.

A SETTLEMENT ESTABLISHED in 1778 at a spot known as Falls of the Ohio grew to become the modern city of Louisville. Today, the Ohio River contains the largest naturally exposed Devonian fossil beds in the world. Transporting travelers across the Ohio, the George Rogers Clark Memorial Bridge has linked Louisville with neighboring Indiana since 1927 (PAGES 16 AND 17).

The Greatest City

© DAN DRY / DAN DRY & ASSOCIATES

Dedicated in July 1999, Louisville's Waterfront Park features some 55 acres of outdoor fun punctuated by the beautiful Ohio River. Officials estimate that as many as 1 million people took advantage of the site's many offerings during its first full year of operation.

DESIGNED BY F.W. MOWBRAY and opened in 1891, Union Station represents one of Louisville's grandest examples of the Richardsonian Romanesque style. Restored in 1979 at a cost of around $2 million, the building is occupied today by the TARC bus system.

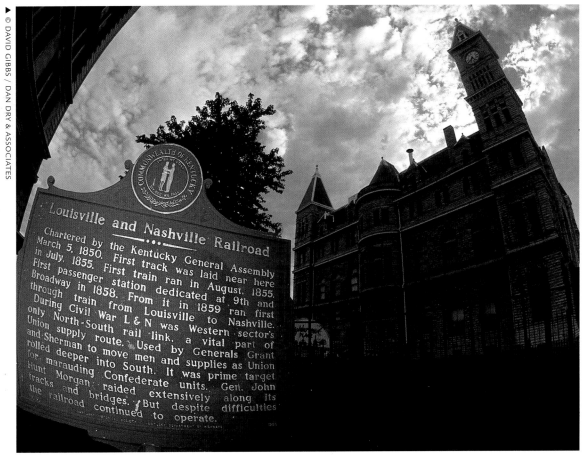

HOME TO A WEALTH OF plays, musical events, and dance performances each year, the Kentucky Center for the Performing Arts opened in 1983 with a gala that included celebrities Diane Sawyer and Lily Tomlin. In addition to its staged events, the center has on permanent display a diverse collection of 20th-century art, featuring works by Alexander Calder, Louise Nevelson, and Joan Miro, among many others.

F ORMED IN 1952, THE Louisville Ballet joins a host of other arts groups to form the core of the city's cultural attractions. Keeping these events alive financially falls, in part, to the Fund for the Arts, which funnels millions of dollars to local organizations.

The historic University of Louisville traces its roots to 1798, when a handful of men formed a seminary that was the state's first school of higher learning. From that inauspicious beginning developed a world-class university with an enrollment of 21,000 students.

Unquestionably one of the best-known of the peaks comprising the Louisville skyline, the pink granite Humana Building dominates its portion of Main Street. Designed by renowned architect Michael Graves, the 27-story skyscraper is home to Louisville-based health care giant Humana Inc.

KNOWN FOR ITS COPPER vats, red wax bottle seals, and unconventional advertising, Maker's Mark joins a long line of distilleries brewing up batches of distinctive Kentucky bourbon. Designated a national landmark in 1980, the Loretto facility is the smallest and oldest distillery currently operating in the nation.

The city doesn't mess around when it comes to decorating for the holiday season. Each year since 1982, Light up Louisville has set downtown ablaze with fireworks and a half million lights.

Each year since 1998, Mayor David L. Armstrong has burned the midnight oil in City Hall, where his plan for Louisville's future reflects more than just machine politics (PAGES 34 AND 35).

NO MATTER WHAT THE SEASON, Louisville residents enjoy a full calendar of activity, with walks in the snow or along the shore during a spectacular sunset ranking high on the list of favorite pastimes.

LONG KNOWN AS AN AREA of great beauty, Louisville is also a mecca of cutting-edge technology. The city serves as home to some of the world's most advanced corporations (PAGES 38-41).

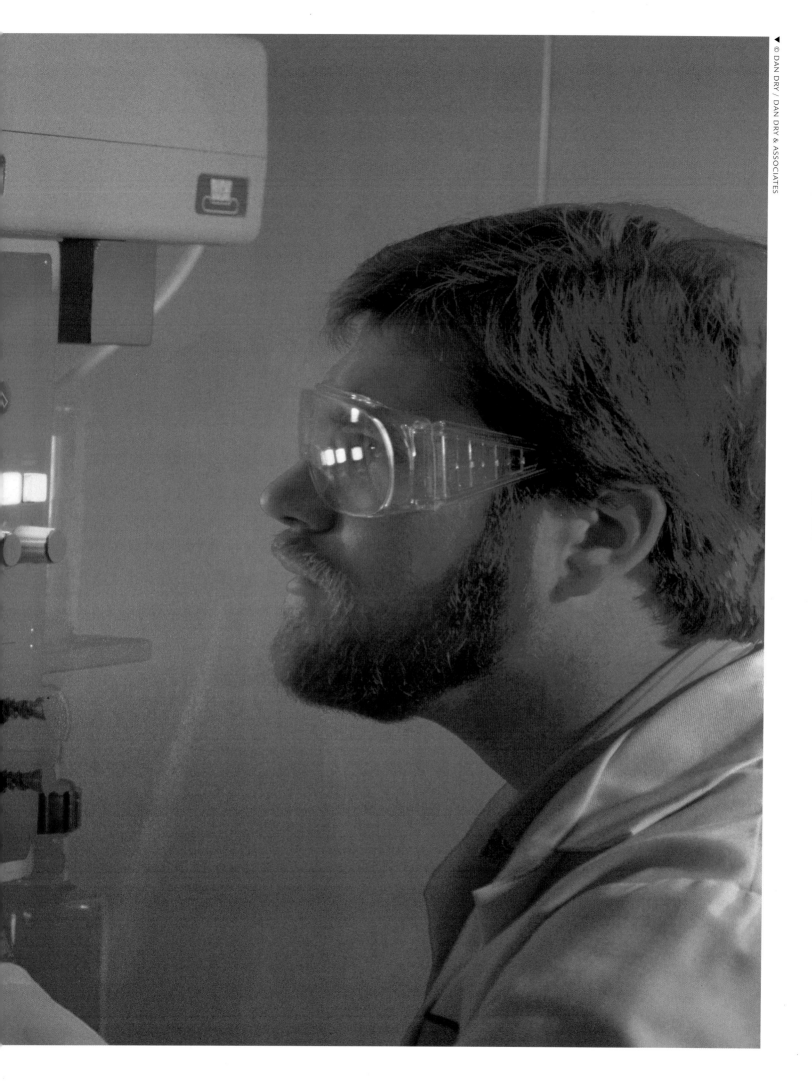

ORGANIZATIONS SUCH AS National Products, Inc. (TOP)—the nation's largest manufacturer of disco balls—and the Louisville Science Center (OPPOSITE) reflect the ingenuity of Louisville residents.

LOUISVILLIANS CAN BE BOTH timely and timeless, depending on such pacesetters as the clock outside the Camberly Brown Hotel (OPPOSITE) and Elizabeth Schaaf of Elizabeth's Timeless Attire (RIGHT).

FEEDING THE SOULS OF Louisville residents with an eclectic dose of food, crafts, and festivals, Lynn Winter of Lynn's Paradise Cafe capitalizes on her quirky sense of style.

A WEALTH OF GOLDEN HUES and dwindling leaves make fall the most colorful of Louisville's seasons.

Leapin' lizards! Sticky situations can sometimes lead Louisville residents to climb the walls.

THE BLUEST SOUTHERN SKY offers a stunning backdrop for Louisville's many sights. Being full of hot air is not always a bad thing, especially if it results in breathtaking views from soaring heights (PAGES 53 THROUGH 55). Particularly fond of hot-air balloons, Louisville boasts several popular events, including the Gaslight Festival and the Great Balloon Race.

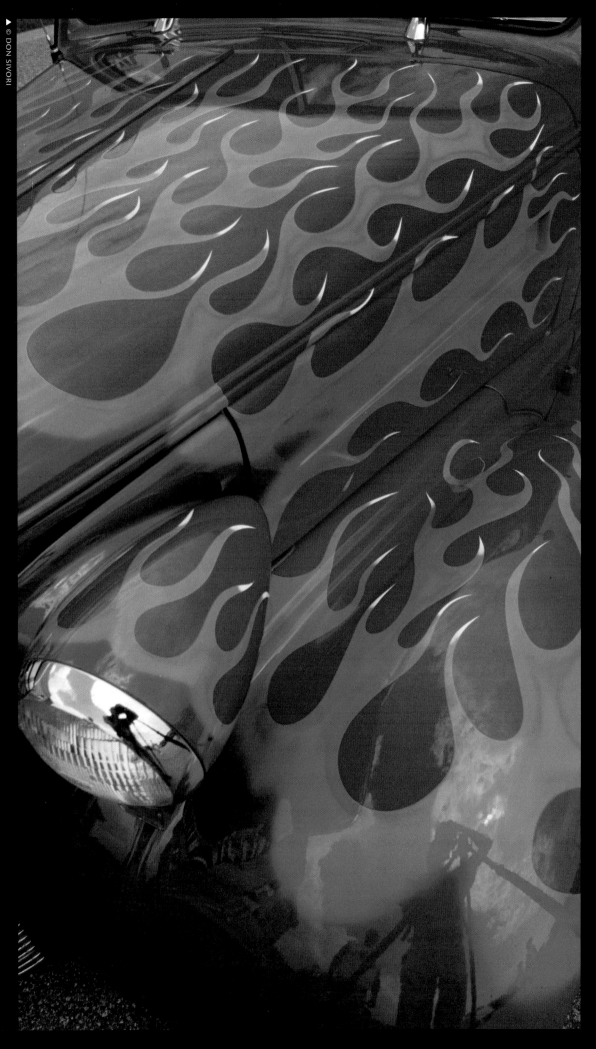

W HETHER AIRBORNE OR land-based, locals get all fired up for some eclectic forms of transportation.

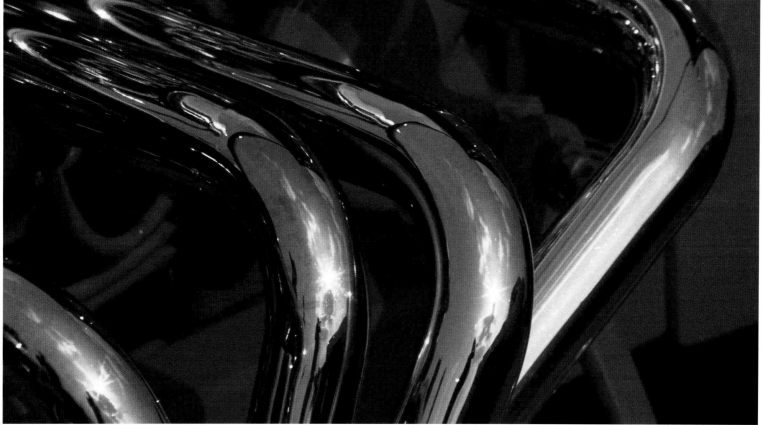

BODIES OF STEEL—WHETHER in the form of classic hot rods or local athletic champions—are nothing new to Louisville. Ringside audiences get a thrill when witnessing the feats of WWF star Rico Constantino (PAGE 60) and boxer Terry Middleton (PAGE 61).

Louisville native and socialite Patricia Barnstable-Brown (OPPOSITE) and African-born entrepreneur Elizabeth Kizito both know how to entertain their hometown. Barnstable-Brown, hostess of one of Kentucky's most glamorous Derby parties, and Kizito, owner of Kizito Cookies, regularly supply residents with a healthy dose of glamour and good cookin'.

As just one of the art galleries on Louisville's East Market Street, Galerie Hertz—named for director Billy Hertz (ABOVE)—represents a revitalization of visual arts in the city. Doing his part to keep the movement alive, local sculptor Ed Hamilton (OPPOSITE)—creator of the nationally recognized African-American Civil War memorial *The Spirit of Freedom* in Washington, D.C.—continues to carve his place in history.

For Louisvillians, self-expression takes a variety of forms. Local tattoo artist Charlie Wheeler (OPPOSITE) creates works of art on human canvases, while those less permanently inclined resort to a clever bumper sticker or 20.

Amid considerable hoopla, magician Laurance Jones and 1976 national hula hoop champ Marie Perry raise performance art in Louisville to an entirely new level (PAGES 68 AND 69).

JEWELER AND SCULPTOR Clifton Nicholson Jr. (ABOVE) draws inspiration for his unique pieces from his beloved surroundings in Scottsburg. Kathy Cary (OPPOSITE), chef/owner of Lilly's in Louisville, transforms garden treats into southern fare with an international flavor.

WITH ITS GIANT BALLOONS, floats, and marching bands, the Pegasus Parade has been a part of the Kentucky Derby Festival since 1956. Each year, an estimated 225,000 people queue up along Broadway to watch the parade pass them by.

MAKING THE TREK FROM Iroquois Park to Market and Sixth streets, some 6,000 athletes participate in the annual Kentucky Derby Festival miniMarathon. The 13.1-mile race is just one of several events leading up to the Kentucky Derby.

W INTER, SPRING, SUMMER, or fall, Louisville delights the senses with a variety of seasonal color (PAGES 76-79).

O**N THE CAMPUS OF THE** University of Louisville, the Alumni Clock Tower (THIS PAGE) marks the hours. Making its appearance on the time flies continuum, the Louisville International Airport (OPPOSITE TOP) handles nearly 4 million passengers annually. But it's the fleeting nature of childhood that shows us the greatest evidence of time's passage.

Located in the old Carter Dry Goods building on West Main Street, the Louisville Science Center (ABOVE) delights both young and old with its interactive displays, IMAX Theatre, and exhibits. Not far away in Waterfront Park, artist Charles Perry's sculpture *Tetra* frames the downtown skyline (OPPOSITE).

The Greatest City

Some residences in Louisville seem palatial enough for a curtsy. From down-home charm to luxurious apartments, the city offers a taste of royalty to everyone.

WHETHER IRONWORK OR stone, the architectural touches that define Louisville appear in both small and large ways. The Gothic spires of First Lutheran Church (PAGES 88 AND 89), a downtown fixture since 1872, tower over East Broadway.

From the stage of the historic Palace Theatre, Bradley L. Boeker (ABOVE RIGHT) gives his regards to Broadway by bringing professional performances to Louisville. Just as the Palace underwent restoration, so do the player pianos and pipe organs lovingly reworked by Bob Martin (OPPOSITE).

MILLIONS OF IMPORTANT packages and documents are routed and rerouted daily in Louisville. The city serves as the main U.S. air hub for United Parcel Service.

DEMURE LOOKS MAY COME and go—unless, of course, you're a statue—but the fashions of Louisville designer Cleah Talbert remain at the forefront of couture.

The Greatest City

OBJECTS OF DESIRE: Some people display them, some people offer advice about them. Actress, model, and Louisville native Catherine McCord (ABOVE) costars on MTV's *Loveline*, a question-and-answer program for the lovelorn. Julie Comer (OPPOSITE) lends her expertise to the art world with exhibitions at her Objects of Desire Gallery.

P UTTIN' ON A HAPPY FACE comes easily for many of the collectibles found in Joe Ley's antique store, as well as for the clients of Alix Adams Models and its owner, Dick Anderson.

COLORFUL AND FLAMBOYANT plumage is not just for peacocks. Local radio personality Coyote Calhoun is almost as well known for his array of costumes as for his on-air antics.

The Greatest City

N ATURAL BEAUTY, WHETHER plant or animal, arrives in vivid color to usher in a time of renewal in Louisville.

© LARRY S. FOSTER

THEY MAY CALL IT PUPPY LOVE, BUT AFFECTION IS ALL around the animal kingdom. From the forlorn gaze of man's best friend to the exotic attractions at the Louisville Zoological Garden, sweetness abounds.

S NOW WHITE IS MORE THAN just a fantasy for Louisville residents (PAGES 106-109). No matter what the season, the area's natural beauty remains undwarfed.

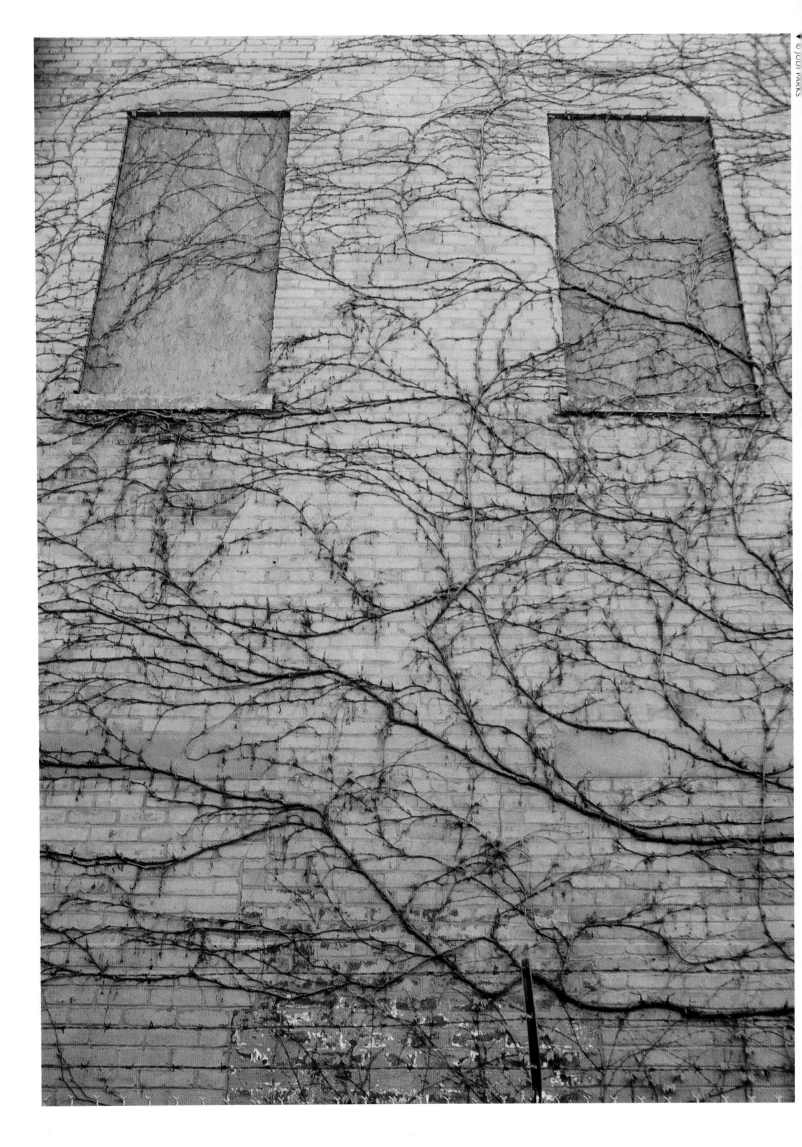

OFFERING THE BEST OF BOTH worlds, Louisville shines—day or night. Rustic on the outskirts, often hectic at its heart, the city reflects both sides of the lifestyle coin.

For a taste of Louisville's musical roots, the soulful sounds of V-Groove (ABOVE), performances by blues harmonica player Lamont Gillispie (OPPOSITE LEFT), or the soothing acoustics of guitarist Pat Kirtley (OPPOSITE RIGHT) cover the spectrum in venues around the city.

A LTHOUGH THEY MARCH TO THE BEAT OF A DIFFERENT drummer, members of a local high school band share common threads with their classic toy soldier counterpart.

Fireworks are an important part of many Louisville celebrations, especially those honoring the nation's war veterans on patriotic holidays.

© LINDA MORTON

MEMORIAL FLAGS ADORN MILITARY TOMBSTONES FOR annual Fourth of July remembrances at Zachary Taylor National Cemetery (ABOVE). The final resting place for numerous veterans, the site is also the burial place of its namesake, the 12th president of the United States.

© JONATHAN POSTAL / TOWERY PUBLISHING, INC.

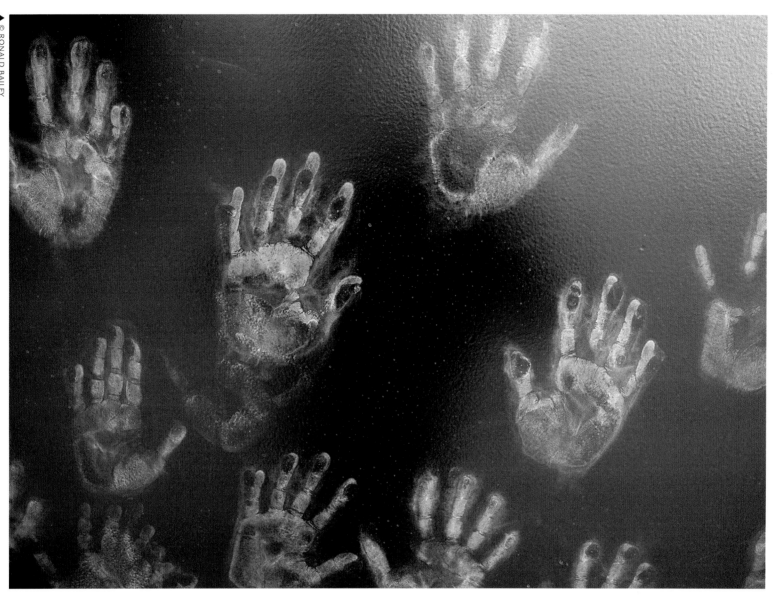

MANY HANDS HAVE TOUCHED the storied history of Louisville. From graveside and other public statuary to the serenity of a soft morning mist in Seneca Park (PAGES 122 AND 123), the city can be eerily beautiful at times.

THERE MAY BE IMITATORS, but few in Louisville can conduct themselves with the flourish of Robert Franz (OPPOSITE), associate director of the Louisville Orchestra and a promising star in the area's arts scene.

Louisville

The fashion sense of Louisville residents can range from the formal styles of yesteryear to the casual wear of today's trendsetters.

SOME LITTLE PIGGIES NEVER have to go to market—especially those that have turned to stone. Across the river in Indiana, however, statues, Grecian columns, and birdbaths from the Concrete Lady do manage to find the way into many of Louisville's yards and gardens.

BATTER UP: Arguably the most recognizable item the city exports, the Louisville Slugger is a mainstay of America's favorite pastime. Each year, the downtown facility generates more than 1.4 million wooden bats—some still handmade for professional ballplayers by specialists such as Danny Luckett (OPPOSITE).

OPENED IN APRIL 2000, Louisville Slugger Field—home to the popular Triple-A Louisville RiverBats—seats some 13,000 fans from its perch on the banks of the Ohio River. A converted rail freight depot serves as the ballpark's front entrance, providing 22,000 square feet of retail shops and restaurants.

THE TOAST OF THE TOWN, Louisville Slugger Field includes 30 private suites, second-level club seating, a continuous concourse around the field, extensive press facilities, concessions and rest rooms, a children's play area, and natural grass on the recessed field.

WHETHER MECHANICAL or mural, fortune tellers and ticket sellers inject an air of nostalgia into daily life.

THE PLAY MAY BE THE THING, but—as with both performance and visual arts—it's all in the execution. Look no further than Squirrelly's Magic Tea Room if a Great Dane named Stormin' Norman Jr. and his magician-master Larry Jones strike your fancy. In the world of live theater, the classic 1901 children's story *Mrs. Wiggs and the Cabbage Patch*, by Louisville native Alice H. Rice, attracted yet another generation of admirers when it arrived on stage in Louisville.

FRAMED IN THE CROSS BEAMS of one of Louisville's three river bridges, the city skyline rises majestically from the banks of the Ohio River.

The inspiration for miniature models and romantic dreams, steamboats occupy a special place in American folklore. The *Belle of Louisville* is the oldest river steamboat still in operation, and, in 1989, was designated a National Historic Landmark.

Remembered for its 400-foot-tall spray, Falls Fountain (RIGHT) no longer plays a featured role in Louisville's waterfront development. Removed in 1998 after breaking down, the $2 million gift to the city faces an uncertain future. Unlikely to fade into the sunset are the city's Ford Motor Co. plants (OPPOSITE). With record production numbers and significant investment from their parent company, the Louisville Assembly Plant and Kentucky Truck Plant combine to produce some 725,000 vehicles annually.

The Greatest City

© DEBORAH BROWNSTEIN

BEAUTIFUL AS WELL AS useful, area fountains offer a much needed respite from the hot summer sun. Outside the new Louisville Water Company building (ABOVE), a decorative glass fountain adds an intriguing accent to the facility's architecture.

© WEASIE GAINES / DAN DRY & ASSOCIATES

Protecting one of Louisville's most precious commodities warrants communitywide effort. Founded locally in 1983, Project Safe Place is a YMCA outreach program designed to help children facing a variety of problems. Businesses displaying the yellow logo offer immediate short-term protection until the appropriate help can arrive.

LONG A PICTURE OF FECUND ferocity, even the savage lion lets its noble guard down now and again.

Not everybody has a head for decorating. But in Louisville, the antique shops and exterior touches scattered about indicate that at least somebody is minding the store.

IN LOUISVILLE, MAKING A pig of yourself is just plain fashionable, especially when visiting the posh Oakroom Restaurant (ABOVE). Located in the renowned Seelbach Hotel, the restaurant specializes in down-home dishes with a contemporary spin. Jo Ross (OPPOSITE) keeps Louisville's fashionably elite up-to-date on contemporary attire through her calendars and newspaper columns.

L OUISVILLE BOASTS A VARIETY of quaint stores offering a vast array of extravagant merchandise. And while the shopping experience can be fulfilling for some, for others it's preferable to sit back and relax.

RESIDENTIAL ARCHITECTURE in Louisville can run the gamut from Victorian to redbrick. Whatever the exterior, the beautiful, old homes found throughout the area form a lovely backdrop for a quiet stroll down a shady street.

© WEASIE GAINES / DAN DRY & ASSOCIATES

SOMETIMES SEEING RED is a good thing, especially when the color adds a brilliant glow to the beautiful sights in Louisville.

The Greatest City

EASY ACCESS TO ABUNDANT family fun—from petting to piloting, and everything in between—is one of Louisville's most attractive features.

For some, mud and water function as sources for good clean—and sometimes dirty—fun. But for the potters at Louisville Stonewear (OPPOSITE), a handful of clay translates into a work of art. And who would have thought the fountain of youth would appear in the form of a big red bucket (PAGES 166 AND 167)?

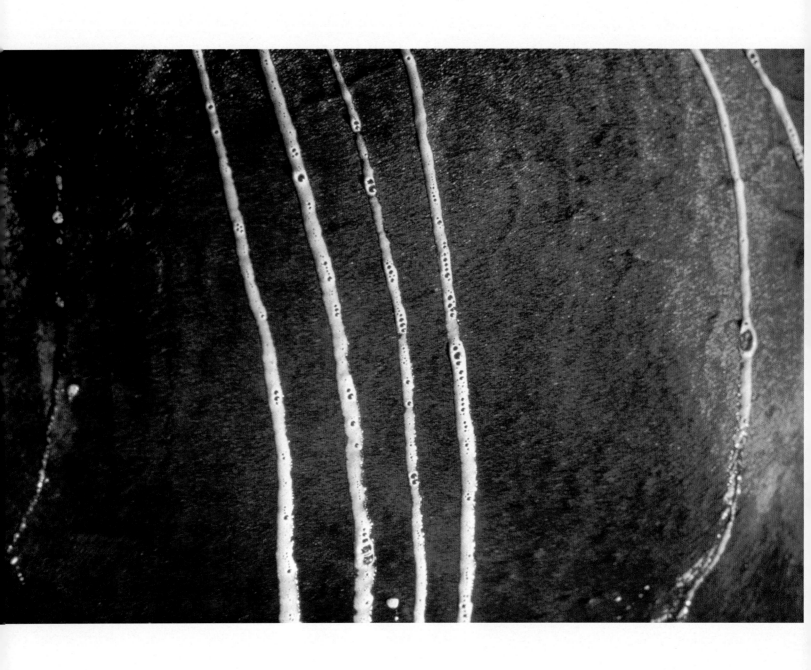

COMPETITION IS FIERCE AT the Louis D. Brandeis School of Law at the University of Louisville (ABOVE), where future lawyers learn how to prevent clients from getting soaked—legally speaking, of course. But you'll get no arguments from the recipient of a nice cooling bath at the annual Kentucky Derby (OPPOSITE).

A FOUNTAIN WITH ARCHES of water 15 feet high dances in the sculpted lighting of Louisville Waterfront Park (ABOVE). Park designer George Hargreaves no doubt got his inspiration from a variety of sources, one of which could have been the natural rays of sunrise across early morning grass (OPPOSITE).

CRAFTSMEN OF DIFFERENT sorts put the swirl in the University of Louisville's University Club and Alumni Center and a stained-glass wind chime at the St. James Court Art Show. The annual three-day event draws a crowd of 300,000 to Old Louisville.

THE NUMEROUS ATTRACTIONS that light up life in Louisville help bring into focus its role as one of the South's premier cities. A particular draw for boaters is the beautiful Ohio River (PAGES 176 AND 177).

LIGHT AND SHADOWS LEND a dramatic air to the campus of the University of Louisville (BOTTOM) and to Churchill Downs (OPPOSITE). But there's nothing quite like the cheeky glow of a youthful face.

No matter what the age, sisterhood—in the flesh or in similar tastes and interests—makes for lasting bonds.

CREATING WORKS OF ART, whether in the food or the furniture field, makes two Louisvillians kindred spirits. Susan Seiller Smith (ABOVE) serves tasty delights at her Bardstown Road restaurant, Jack Fry's. Unique creations by furniture maker/artist Ted Harlan (OPPOSITE) have made their way to museums worldwide, including the Smithsonian Institute.

Consider the cigar. Once acceptable only in the domain of lush and private men's clubs, these days the stogie transcends gender.

WHERE THERE'S SMOKE, there's fire, and where there's fire, there's sure to be a Louisville firefighter on hand to douse it.

DEDICATED TO THE
MEMORY OF THOSE
LAW ENFORCEMENT
OFFICERS WHO HAVE
GIVEN THEIR LIVES IN
THE LINE OF DUTY

The Fallen Firefighters Memorial (BOTTOM) and the Law Enforcement Officers Memorial (OPPOSITE BOTTOM) both occupy special places in downtown's Jefferson Square. Renowned Louisville sculptor Barney Bright created the monument to firefighters, dedicated in 1996.

VISIBLE FROM BLOCKS AWAY, the six-story-tall replica of the bat that made Louisville famous rests against the entrance to the Louisville Slugger Museum (PAGE 190). The Hillerich & Bradsby Co. made its first bat in 1884, christening it after its home city 10 years later. Today, the company's headquarters houses the museum's baseball-related attractions, including interactive displays, memorabilia, and a tour of the complete bat-making process—from white ash to home run contender (PAGES 191-193).

The Greatest City

FLYING IN FORMATION IS nothing new, either for ducks or for Navy jets. Giving flight to the health care hopes of Louisville-area youngsters falls to Kosair Children's Hospital (PAGE 196), the region's only full-service facility exclusively for kids.

PUSHIN' AND PULLIN': Defying the laws of physics sometimes takes little more than a pint-sized show of brute force. For several civic-minded teams participating in the UPS Plane Pull (OPPOSITE), it was a giant-sized heart that won the day. The event was a fund-raiser for Special Olympics Kentucky.

© WEASIE GAINES / DAN DRY & ASSOCIATES

N EVER LET IT BE SAID THAT there isn't plenty to keep locals busy. Louisville's many outdoor paradises give residents a natural high.

O ver the years, bikers have managed to shed their outlaw image and turn the activity into a family affair. Even the tykes get into the spirit—training wheels and all.

SOME 270,000 PEOPLE CALL Louisville home, and their faces reflect a city of vigor, vitality, and variety.

IN ADDITION TO ITS HISTORIC inner city neighborhoods, Louisville offers a wealth of communities on its outskirts. When lightning strikes—at least in the form of human emotion—can a home in the burbs be far behind?

Ascending the climbing wall at Rock Sports can be a, well, hair-raising experience. Maureen Kunz (OPPOSITE) has clearly accepted the Louisville business' motto challenge: We Double Dog Dare Ya!

Under the leadership of Coach Denny Crum since 1971, the University of Louisville Cardinals basketball team has scored a slam dunk with fans and pollsters for years. Crum has taken the team to 23 NCAA tournament appearances, going home with championship trophies on two of those occasions.

COME FALL, FOLKS IN Louisville start flippin' out over University of Louisville football. Under the guidance of Coach John L. Smith, the team competes in Conference USA.

It takes all kinds of music to keep the city's collective toes tapping. One local band that's developing a loyal following is My Morning Jacket (OPPOSITE BOTTOM), known for its hard-to-define, distinctive sound. But there's nothing undefinable about the sounds of the blues, taking main stage annually at the Louisville Blues Festival (THIS PAGE).

A FIXTURE AT ALPINE ICE Arena on Gardiner Lane for some 30 years, senior instructor Lydia Herron (ABOVE) coaches Louisville's synchronized skating team, Moms on Ice.

Equally familiar to the city's audiences, Helen Starr (OPPOSITE) performed with the Louisville Ballet for 22 years. Today, she keeps her toes in the company's pirouettes working with husband

Alun Jones, the group's artistic director. Performers demonstrating sculptured movements from India reflect the diversity of the area's cultural interests (PAGES 216-217).

CASTING ONE'S LOT IN LIFE sometimes means just going with the flow (PAGES 220-223). But if an afternoon of fishing doesn't qualify as an occupation in your book, the University of Louisville College of Business and Public Administration might be more appropriate (OPPOSITE). The department's modern facility puts its students in touch with the latest in business technology.

KIDS AND PARENTS ALIKE have embraced the variety of exotic creatures featured at the Louisville Zoological Gardens. Under the direction of William Foster (OPPOSITE), the 75-acre site plays host to around 700,000 people annually.

Folks who don't really give a hoot when it comes to high fashion better stay away from the Kentucky Derby, where the horses sometimes take a back seat to what's in vogue (TOP). Friendship, however, never goes out of style—always providing bosom buddies something to croon about.

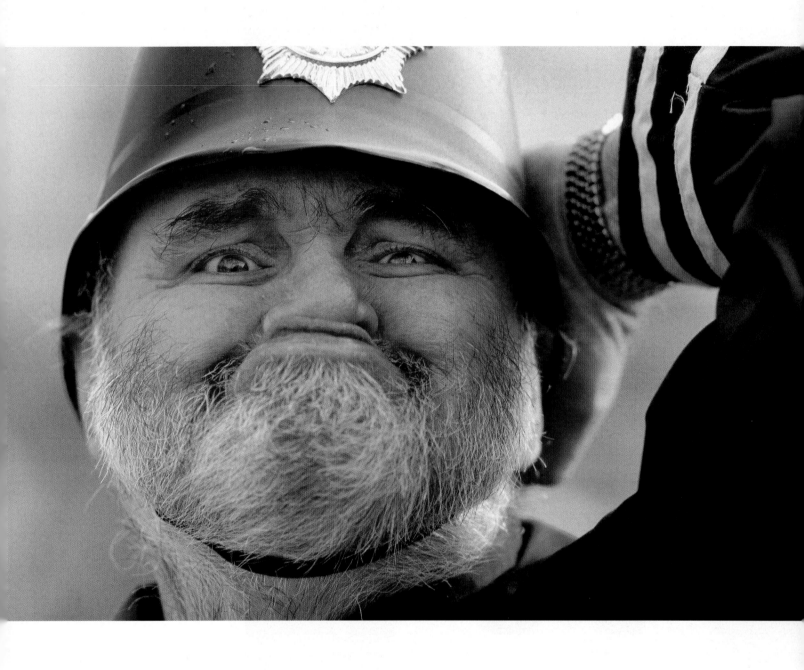

To local residents, a foal can become the stuff that dreams are made of. In addition to the area's excellent equine crop, its Homo sapiens population displays a variety of talents as well.

Face it: No matter how good the makeup job or the artist, you just can't compete with the real thing when it comes to fierceness (PAGES 230-233).

OUR LITTLE
EDDIE

He has gone to his Heavenly home
His favorite Dog still watching o'er his tomb

T.F. EVANS

© BOB HELVEY

TRIBUTES TO THE DEPARTED—BE THEY ARTISTICALLY elaborate or touchingly simple—evoke thousands of memories for those who are left behind.

Since Squire Boone, Daniel's brother, reportedly traveled here in 1781 to preach the area's first sermon, religion has been a part of Louisville's daily life. Protestant, Catholic, Jewish, and other faiths congregate to worship in the city that is headquarters for the Presbyterian Church USA and home to the Southern Baptist Theological Seminary.

© JONATHAN POSTAL / TOWERY PUBLISHING, INC.

© JONATHAN POSTAL / TOWERY PUBLISHING, INC.

INSPIRATION COMES IN ALL forms, from the spiritual to the physical. Ballerinas at the Louisville Ballet School find an outlet for expression in the world of dance. Attendees at Celebration 2000 reached for a divine source to guide them through the new millennium.

© JONATHAN POSTAL / TOWERY PUBLISHING, INC.

THE STRENGTH OF EMOTION often serves as a driving force—albeit for a wide spectrum of activities. Some 5,000 people gathered in Louisville Gardens for St. Stephen Baptist Church's Celebration 2000 on December 31, 1999 (ABOVE LEFT AND OPPOSITE LEFT). The facility draws a somewhat different crowd for bouts of Ohio Valley Wrestling, a training organization for the World Wrestling Federation.

FROM THE DEW-DAMP WEAVE of a spider's web, to the dynamic flare of fireworks over the city, to an energetic burst of lightning (PAGES 246-247), Louisville sparkles with energy.

© DON SIVORI

AUTHORITIES BLAMED VANdalism for the near sinking of the *Belle of Louisville* on August 24, 1997. That same year, Mother Nature and some high floods put the Louisville Water Tower in jeopardy. Despite these setbacks, the gray skies cleared up over the city (PAGES 250 AND 251).

© DAN DRY / DAN DRY & ASSOCIATES

The quiet moments at Churchill Downs—made all the prettier by a winter snow—belie the frenzy of activity that erupts when racing gets underway.

© RONALD L. PROFUMO

A AAAAANNNND, THEY'RE off. Each spring and fall, Churchill Downs hosts Thoroughbred horse racing at its best. Come May, a capacity crowd of 50,000 spectators gathers for the annual Run for the Roses— the Kentucky Derby. Since 1875, the world's top three-year-olds have run neck-and-neck toward the finish line of the one-mile track (PAGES 254-257).

© WEASIE GAINES / DAN DRY & ASSOCIATES

© WEASIE GAINES / DAN DRY & ASSOCIATES

© DAN DRY / DAN DRY & ASSOCIATES

The Greatest City

255

© DAN DRY / DAN DRY & ASSOCIATES

Taking a peek into a pumpkin—or a cool drink of water fresh from the hose—constitutes fun times for a couple of Louisville kids. Adults, however, exhibit slightly differing expectations when it comes to entertainment. A Derby party, circa the 1970s (TOP), and its modern day counterpart (OPPOSITE RIGHT) show that the more things change, the more they stay the same.

GRAVITY AND SOUND CONNIVE to create a rambunctious waterfall at Louisville's Great Lawn in Waterfront Park. The fluid commotion flows in vivid contrast to the pristine blanket of snow at another of the city's well-known attractions—Churchill Downs.

Thanks to Diaper Dip classes at the YMCA, Louisville area kids learn not to fear the water. Similar early training must have paid off for Tori Murden-McClure (OPPOSITE), the Louisville resident who gained worldwide attention in December 1999 as the first woman and first American to row solo across the Atlantic Ocean.

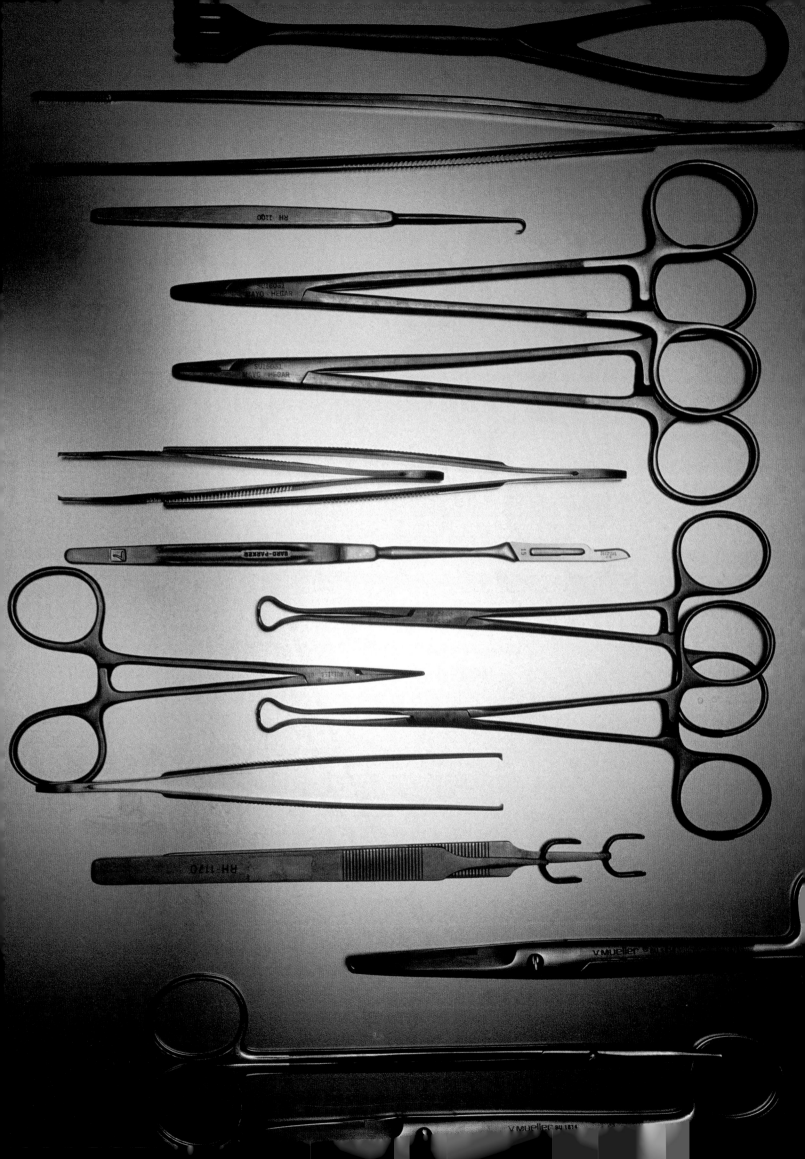

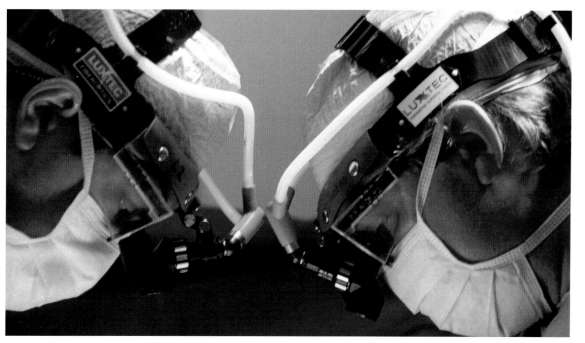

SINCE THE EARLY 1800S, when malaria swept through the area, doctors and the health care industry have played a role in developing Louisville. The city's medical facilities are known around the world, particularly for advances in heart transplants in affiliation with the Humana Heart Institute International.

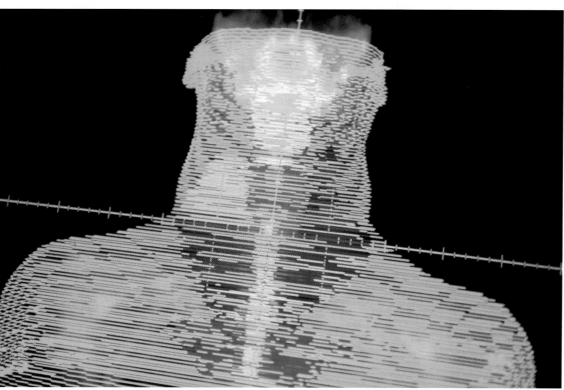

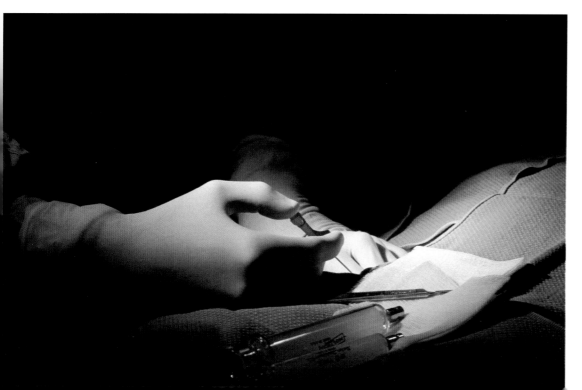

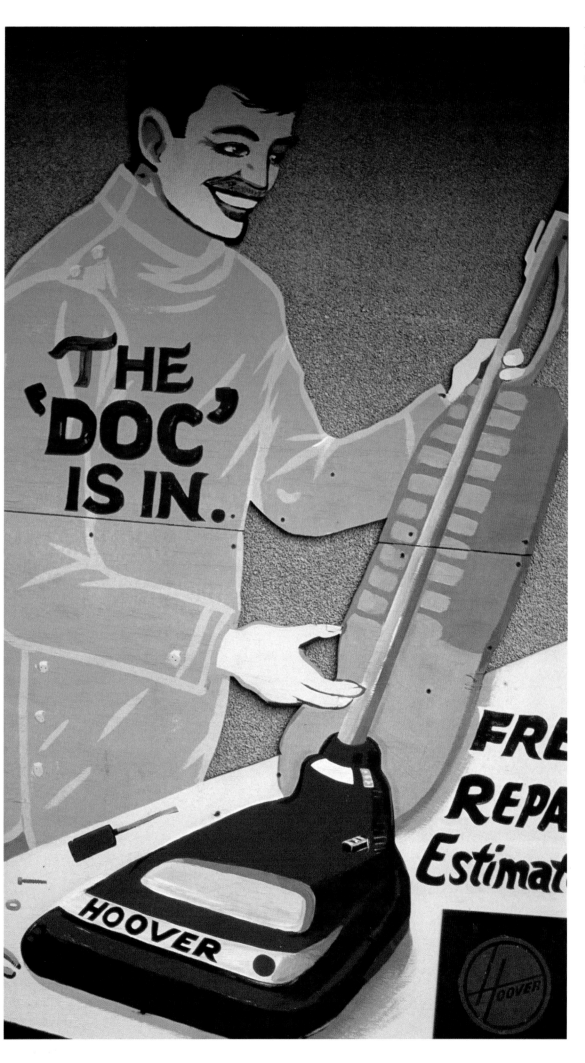

LOUISVILLIANS CAN BOAST A cure for just about anything—whether civil, medical, or mechanical.

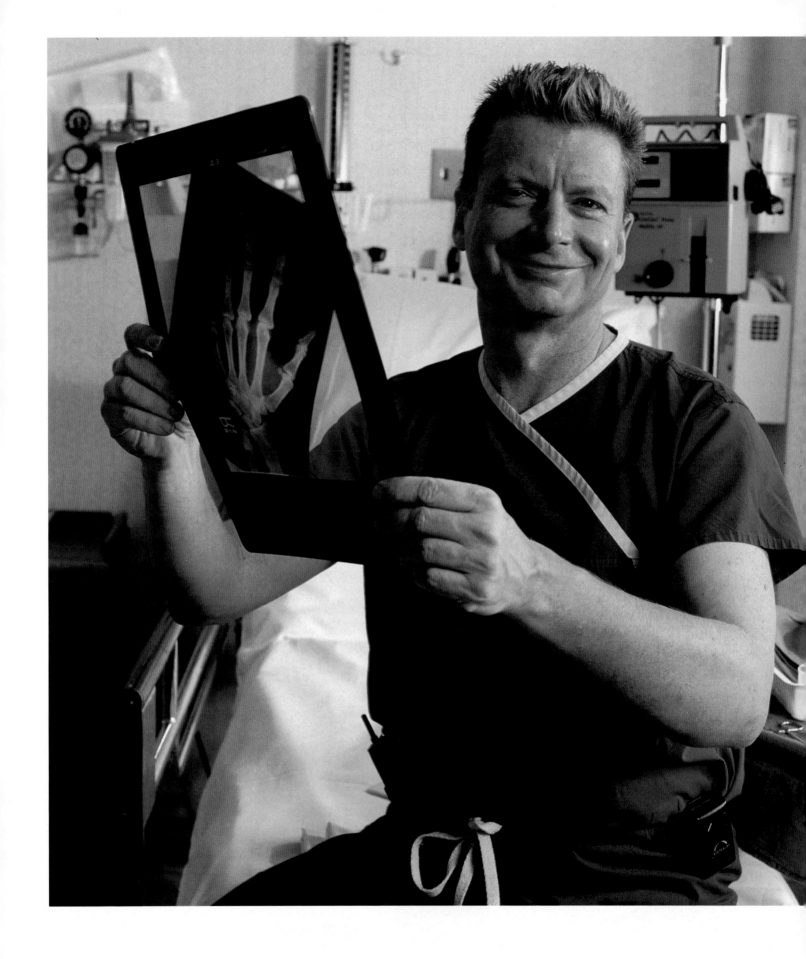

A WEALTH OF TALENT KEEPS Louisville going. Architect Ted Bressoud (ABOVE) lends his personal flair to the city's businesses and homes. Dr. Warren C. Breidenbech, of Kleinert, Kutz and Associates Hand Care Center (OPPOSITE), led a team of Louisville doctors through the first hand transplant in the United States.

THE STILL WATERS OF THE OHIO River at night offer Louisvillians a scenic respite from the responsibilities and pressures of a hectic workday.

AS THE SUN RETIRES FOR another day, Louisville residents can relax, knowing they will wake up in one of the world's greatest cities.

© DAN DRY / DAN DRY & ASSOCIATES

© DON SIVORI

Profiles in Excellence

A LOOK AT THE CORPORATIONS, BUSINESSES, PROFESSIONAL GROUPS, AND COMMUNITY SERVICE ORGANIZATIONS THAT HAVE MADE THIS BOOK POSSIBLE. THEIR STORIES—OFFERING AN INFORMAL CHRONICLE OF THE LOCAL BUSINESS COMMUNITY—ARE ARRANGED ACCORDING TO THE DATE THEY WERE ESTABLISHED IN THE LOUISVILLE AREA.

AAF International ■ Adelphia Business Solutions ■ ADVENT ■ Aegon Insurance Group ■ Anthem Blue Cross and Blue Shield in Kentucky ■ Bakery Chef, Inc. ■ Bellarmine University ■ Brown, Todd & Heyburn PLLC ■ Business First ■ Cambridge Construction Company ■ Carlson Wagonlit Travel/WTS ■ Cintas Corporation ■ Classroom Teachers Federal Credit Union ■ Clear Channel Communications, Inc. ■ Dismas Charities, Inc. ■ Fire King International Inc. ■ Ford Motor Company ■ GE Appliances ■ Goldberg & Simpson, PSC ■ Gordon Insurance Group ■ Greater Louisville Inc. ■ Heick, Hester & Associates ■ Humana ■ Infinity Outdoor ■ Insight Communications ■ Jewish Hospital HealthCare Services ■ Johnson Controls ■ Kentucky Fair and Exposition Center/Kentucky International Convention Center ■ The Kroger Company ■ L&N Federal Credit Union ■ LG&E Energy ■ LabCorp ■ Lear Corporation ■ Lightyear ■ Louisville and Jefferson County Convention & Visitors Bureau ■ Louisville/Jefferson County Metropolitan Sewer District ■ Manpower Inc. ■ Micro Computer Solutions ■ Muhammad Ali Center ■ Musselman Hotels ■ NTS Development Company ■ Neace Lukens ■ Norton Healthcare ■ OPM Services, Inc. ■ Papa John's International ■ Park Federal Credit Union ■ Peter Built Homes, Inc. ■ Presbyterian Church (USA) ■ PRIMCO Capital Management ■ Prudential Parks & Weisberg Realtors* ■ Roman Catholic Archdiocese of Louisville ■ St. Xavier High School ■ Samtec, Inc. ■ Steel Technologies Inc. ■ Süd-Chemie Inc. ■ The Sullivan Colleges System ■ Sun Properties ■ SYSCO ■ TechRepublic ■ Thomas Industries Inc. ■ Torbitt & Castleman ■ Trinity High School ■ United Parcel Service (UPS) ■ Vincenzo's ■ WDRB Fox 41 ■ Wyatt, Tarrant & Combs ■

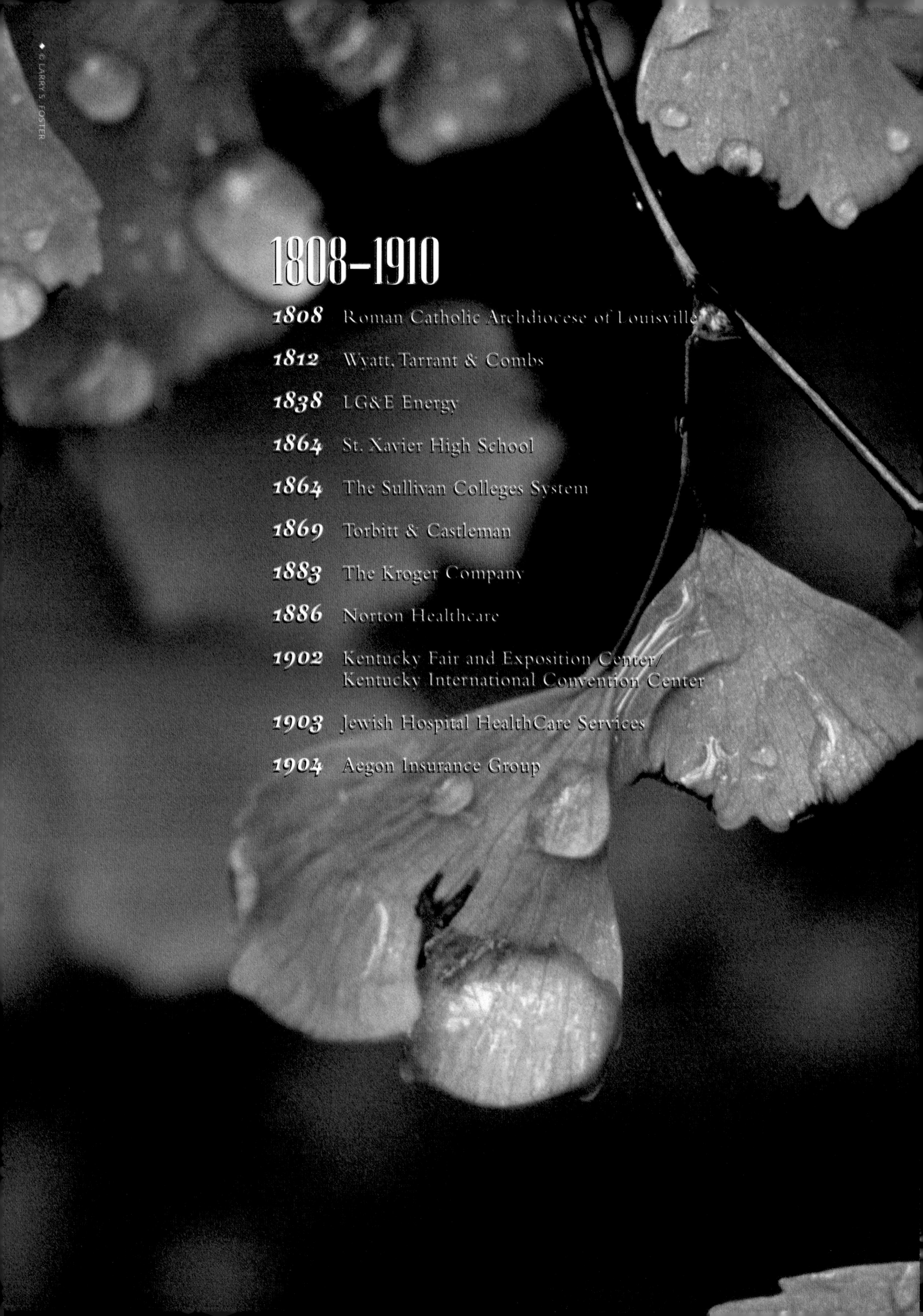

© LARRY S. FOSTER

1808–1910

1808	Roman Catholic Archdiocese of Louisville
1812	Wyatt, Tarrant & Combs
1838	LG&E Energy
1864	St. Xavier High School
1864	The Sullivan Colleges System
1869	Torbitt & Castleman
1883	The Kroger Company
1886	Norton Healthcare
1902	Kentucky Fair and Exposition Center/ Kentucky International Convention Center
1903	Jewish Hospital HealthCare Services
1904	Aegon Insurance Group

Roman Catholic Archdiocese of Louisville

The Roman Catholic Archdiocese of Louisville enters the new millennium strongly anchored in a rich past as one of the earliest dioceses in the United States. After escaping religious oppression in England and France, early Catholic immigrants found their way to Kentucky and settled there during the late 1700s. Although this frontier state seemed an unlikely location, it would one day become one of the larger dioceses in the nation. In 1808, the Diocese of Bardstown was established, comprising a huge area including the present-day states of Kentucky, Indiana, Illinois, and Ohio. This diocese would eventually give birth to 44 new dioceses and archdioceses. By 1841, the Vatican had moved the seat of the diocese to the rapidly growing port town of Louisville. In 1937, Louisville became an archdiocese.

Great Leaders of the Past

The Reverend Stephen Badin, known as the "apostle of Kentucky," came to the area in 1793, and a regular congregation began meeting in the decade that followed. At that time, Badin's parish was known as St. Louis Church. It was a humble beginning of what would become the Cathedral of the Assumption.

The first bishop of the Diocese of Bardstown was Benedict Joseph Flaget, who arrived as bishop for the diocese in 1811. Flaget had three major priorities for this frontier diocese: build a seminary, establish a religious order committed to service and education, and build a cathedral.

During his more than 40-year tenure as bishop, Flaget worked diligently on these goals. By 1811, he had opened St. Thomas Seminary, and in 1812, three women came forward to take religious vows as Sisters of Charity of Nazareth. Both the seminary and the establishment of the Sisters of Charity helped fulfill one of Flaget's primary concerns: education. He achieved the third goal with the opening of a cathedral in Bardstown in 1816. After the seat was moved to Louisville in 1841, Flaget began work on the second phase of this goal—the building of the Louisville Cathedral. He died in 1852, just before the cathedral was completed.

Another prominent Catholic who called the Archdiocese of Louisville home was Thomas Merton (1915-1968), a Kentucky Trappist monk. Merton was one of the most influential figures in 20th-century American Catholic history. His autobiography, *The Seven Storey Mountain*, sold more than 1 million copies and has been translated into 28 languages.

Merton wrote more than 60 other books and hundreds of poems and articles on topics ranging from monastic spirituality to social justice.

First built in 1849, the Cathedral of the Assumption is the spiritual center of the archdiocese. It serves as a spiritual, civic, and intellectual center for the broader community.

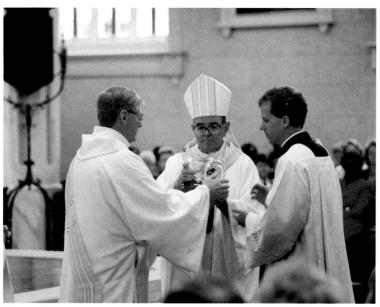

ARCHBISHOP THOMAS C. KELLY, O.P., PRESIDES AT THE REDEDICATION OF THE CATHEDRAL AFTER A MAJOR RENOVATION THAT WAS COMPLETED IN 1994.

THE CATHOLIC CHURCH OF TODAY

Now the primary diocese out of four in Kentucky, the Archdiocese of Louisville extends through 24 counties in Central Kentucky, from the northernmost counties of Trimble and Henry to Monroe, Cumberland, and Clinton at the Tennessee border. One hundred twenty-four parishes, 69 elementary and secondary schools, three colleges and universities, 20 diocesan agencies, and countless other Catholic organizations serve 200,000 Catholics within the archdiocese.

There is great diversity within the Archdiocese of Louisville. Within the 24 counties, Catholics live in urban areas, farm communities, and small towns. Catholics exist in some part of the archdiocese as members of one of the majority faith traditions and in other parts as a tiny minority. In addition to the many Catholics of European descent, there are Catholics of African, Vietnamese, Hispanic, Filipino, Korean, Haitian, and other heritages.

Led by Archbishop Thomas C. Kelly, O.P., the priorities of the diocese flow from its mission to proclaim the good news of Jesus Christ. The archdiocese, together with its parishes, schools, agencies, and other organizations, responds to religious, educational, and social service needs of people from all faith traditions throughout Central Kentucky. The multitude of programs and ministries sponsored by the archdiocese all are intended to proclaim the Christian message in word and action.

The archdiocese serves as a central support system for parishes, schools, families, and individuals. Twenty archdiocesan agencies support the work of parishes and schools through direct financial support, such as tuition assistance and insurance for employees, and through training of parish staff and volunteers. The archdiocesan agencies do things that no one parish could accomplish, including advocacy work through the Catholic Conference of Kentucky, a vast array of social services through Catholic Charities, and communication to every Catholic household through the archdiocesan newspaper, the *Record*, and through television and radio programming.

Kelly's emphasis is on service. "Why do we exist? We exist to serve, following in Christ's footsteps and inviting all we meet to know and love Christ as we know and love Him," he says. "We share with all Christians throughout the world a firm belief that the good news of Jesus Christ speaks directly to the highest hopes and deepest sorrows of the people of this region, and directly to the human needs of people and communities in every time and place."

CATHOLIC SCHOOL EDUCATION HAS BEEN AN INTEGRAL PART OF THE MINISTRY OF THE ARCHDIOCESE SINCE ITS BEGINNING. TODAY, CATHOLIC ELEMENTARY SCHOOLS, SECONDARY SCHOOLS, AND COLLEGES LOCATED WITHIN THE ARCHDIOCESE SERVE MORE THAN 36,000 STUDENTS.

WYATT, TARRANT & COMBS

For nearly 200 years, Wyatt, Tarrant & Combs has helped shape the Greater Louisville area. The law firm traces its roots back to William Christian Bullitt, a nephew of Patrick Henry, who offered the simple legal services required for the frontier life of Louisville in 1812. Since that time, the firm has grown substantially in both size and status. Throughout the years, several members of the firm have held prestigious leadership positions, including the offices of governor of Kentucky, congressman, state legislator, and mayor of Louisville.

What began as a small firm offering basic legal advice is today one of the region's largest law firms with more than 230 attorneys in seven offices located throughout Kentucky, Indiana, and Tennessee. Its professional staff offers resources and experience rivaling those found in the nation's largest law firms. The teams at Wyatt, Tarrant & Combs offer the personalized service that clients have come to expect of smaller, specialized firms, and yet the firm has a comprehensive practice that incorporates all aspects of business and litigation expertise.

A Talented Team

Regardless of a firm's size, its level of quality still depends on its individual lawyers and their ability to work as a team. Managing Partner Stewart Conner believes Wyatt, Tarrant & Combs excels because of its team of talented lawyers and their ability to work together. "The most unique thing about our firm is that, despite our growth, we have managed to maintain a very collegial atmosphere," says Conner. "From the youngest associates to the most senior partners, we all support each other. That network of support strengthens us as a team and strengthens our representation of clients in every situation."

According to Virginia Hamilton Snell, partner and chair of Wyatt, Tarrant & Combs' professional personnel recruiting committee, that collegial atmosphere is a real selling point. "When we recruit associates, our team spirit and workforce diversity attract them immediately," she says. Snell adds that diversity among the firm's management is also important.

And for J. Michael Brown, that atmosphere played an important role in his decision to leave his position as director of law for the City of Louisville to become a partner at Wyatt, Tarrant & Combs. "I made sure that this firm was somewhere where I could be myself," says Brown. "My experience was mainly in small firms and I thought a large firm would be too stuffy, but Wyatt has been a great environment."

In fact, Brown says the size of the firm is an asset for its clients. "Working in such a large group has really been advantageous for me and my clients. At Wyatt, Tarrant & Combs, when a client hires one lawyer, he or she is, in effect, getting the experience of more than 200," says Brown. "If I have a question about some area of law that I am not extremely familiar with, there is someone in this firm who contributes his or her expertise."

"In our global economy, having a large group of experts to rely on will become increasingly important," says Jeffrey E. Wallace, chair of the firm's international trade practice group. "Louisville is destined to become a significant center for E-commerce, a business which has no geographic boundaries, as a result of our distribution hubs. Businesses will have to rely on law firms with experience

Wyatt, Tarrant & Combs' leadership has fostered an atmosphere of collegiality and genuine appreciation for clients' needs.

WYATT, TARRANT & COMBS HAS BEEN PROFILED IN *America's Greatest Places to Work with a Law Degree.*

in a multitude of international trade disciplines. Our firm has many members who not only are familiar with those disciplines, but also have lived and worked abroad. That is a real advantage."

Clients seek attorneys who can see the big picture. Instead of relying on law firms only when they have a legal problem, businesses view law firms as part of the process to help them implement sound strategies. "We become a crucial part of the client's team. We don't just sit back and offer advice; we work with each client to help their business advance," says J. Larry Cashen, chair of Wyatt, Tarrant & Combs' health care practice group.

GIVING BACK TO THE COMMUNITY

All lawyers at Wyatt, Tarrant & Combs are required to provide a certain amount of pro bono work for the community, but the firm's attorneys go far beyond just the required pro bono work. "Our commitment to the community can be traced back to Wilson Wyatt, who expected every attorney in the firm to give back and participate in community activities. That spirit remains today in our firm," says Grover C. Potts Jr., a member of the firm's operating committee. As someone who takes the directive of community involvement seriously, Potts serves as the chairman of the board for the Presbyterian Community Center. So far, under Potts' leadership, the center has raised $4 million to be used for the completion of a community center serving the residents of the economically depressed Shelby Park community.

"Our attorneys are involved in every kind of community activity imaginable, including serving on the boards of arts groups, public service agencies, and religious organizations," says Conner. "The commitment to leadership—in our community and in our profession—is a thread that runs through the entire fabric of the firm."

THE FIRM WAS ESTABLISHED BY DISTINGUISHED ATTORNEYS (FROM LEFT) JOHN E. TARRANT, WILSON W. WYATT, SR., AND BERT T. COMBS.

LG&E Energy

LG&E Energy is a Fortune 500 company that provides diversified energy services—including power generation and project development, retail gas and electricity services, and asset-based energy marketing—to national and international markets from its Louisville headquarters. The company serves more than 1 million retail gas and electric customers, and employs more than 5,000 people.

From left:

LG&E Energy provides low-cost, reliable electricity to customers in nearly 100 Kentucky counties through its subsidiaries Louisville Gas and Electric Company (LG&E), Kentucky Utilities Company (KU), and Western Kentucky Energy (WKE).

LG&E Energy employees work hard to provide customers with exceptional service.

LG&E Energy traces its roots back to 1838, when it was called the Louisville Gas and Water Company. The firm helped install gas pipes locally, and on Christmas Day, 1839, Louisville become the first city west of the Allegheny Mountains—and the fifth in the United States—to have gaslights for its streets. LG&E Energy's tradition of innovation and forward-looking strategies has continued to this day, helping it to become one of the nation's most admired energy companies.

Today, LG&E Energy owns and operates Louisville Gas and Electric Company (LG&E), an electricity and gas utility that serves Louisville and 16 surrounding counties; Kentucky Utilities Company (KU), an electricity utility serving 77 Kentucky counties and five counties in Virginia; and Western Kentucky Energy (WKE), a nonutility electric generating business. With a total generation capacity of 7,300 megawatts, these businesses serve some 880,000 electricity customers and about 290,000 gas customers over a distribution network that covers approximately 27,000 square miles. LG&E Energy also owns and operates nonutility power plants in six states and in Spain, owns interest in three natural gas distribution companies in Argentina, and owns CRC-Evans Pipeline International, Inc. In total, LG&E Energy controls a power-generating capacity of nearly 9,000 megawatts.

A Commitment to the Environment and the Community

LG&E Energy's dedication to providing top-quality energy services translates into a strong commitment to protecting the environment and to supporting the communities it serves. Proactive in environmental protection, LG&E Energy has enacted policies ahead of federal regulations concerning the environment. In the early 1970s, for example, the company began a pilot project to use scrubbers, or air pollution control devices, to reduce sulfur dioxide emissions from its power plants. LG&E Energy was one of the first utility companies in the nation to use these scrubbers. This program was so successful that the U.S. Environmental Protection Agency redesignated Jefferson County as an Attainment Area in the 1980s. In addition, the company actively

LG&E Energy has a number of wind plants throughout the United States. The company's Texas Wind Plant is the largest wind plant in the state of Texas—and the largest in the United States, outside of California (top).

LG&E Energy owns unique underground storage fields that enable the company to keep natural gas costs low (bottom).

participates in the local Ozone Prevention Coalition, which promotes ride sharing on Ozone Action Days—hot, humid summer days.

Equally important to the company is education in the community. In 1989, LG&E Energy partnered with Jefferson County Public Schools to develop the Environmental Tech Prep Program at Valley High School. This program enables students to gain an awareness of environmental issues and to obtain the necessary skills for employment in an environmental discipline, or the academic background to pursue a postsecondary degree. LG&E Energy provides financial assistance for the program, as well as classroom instruction and curriculum development.

The company also supports many other educational programs and community organizations, including an environmental institute for teachers. This annual, three-day program gives teachers an opportunity to visit a power plant, a landfill, a recycling center, water and wastewater treatment plants, and a chemical production facility. The teachers can then incorporate these experiences into their curricula to help students become environmentally aware.

Facing the Challenges of the Future

LG&E constantly strives to maintain its role as one of the nation's largest low-cost energy provider firms. The company has earned national attention, and has received many prestigious awards and honors in recognition of customer service.

On February 28, 2000, LG&E Energy announced a definitive merger agreement with PowerGen plc of the United Kingdom. This transaction offered an immediate opportunity to provide significant value to shareholders, employees, customers, and the Commonwealth of Kentucky. The firm entered into this transaction to optimize the value of one of the country's most efficient energy services companies at a time when the energy industry is struggling to take command of an increasingly competitive marketplace.

Whatever changes may occur, LG&E Energy will continue to be a forward-looking, environmentally proactive, community-involved company seeking to provide diversified, low-cost energy services for its customers in both the national and the international markets.

St. Xavier High School

Since 1864, St. Xavier High School has been educating young men in an atmosphere that provides outstanding academics and quality moral values. Founded by the Xaverian Brothers, St. Xavier offers a private, Catholic secondary education for young men. ■ Although the school has changed locations and experienced phenomenal growth throughout its history, St. Xavier's basic philosophy has changed little. The guiding principles of the Xaverian Brothers continue to serve as the foundation of St. Xavier today. In fact, it is one of only 12 schools nationwide still sponsored by the Xaverian Brothers.

As a Catholic school, St. Xavier takes seriously its role as an extension of the teaching ministry of the Catholic Church, providing many opportunities for the religious and moral formation of its students. All aspects of the school's programs—from the board of directors level through the boys' classroom experiences—are guided by the principles of the Xaverian Brothers and the beliefs of the Roman Catholic Church.

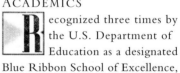

A diverse group of young men from all walks of life from throughout the Greater Louisville region are attracted to St. Xavier High School. An integral part of the school's philosophy is the commitment to offer the opportunities of the St. Xavier spirit, tradition, and experience to all students regardless of financial means. More than 25 percent of the student population receives tuition assistance.

Award-Winning Academics

Recognized three times by the U.S. Department of Education as a designated Blue Ribbon School of Excellence, St. Xavier is well known for the quality of its academic programs. Structured as a college preparatory high school, St. Xavier provides a strong liberal arts curriculum that supports the development of the complete person.

"We take students from where they are, and push them as far as they can go academically," says Dr. Perry Sangalli, principal. Via three different academic levels—honors, academic, and traditional—students are able to enter a program of study that challenges their intellectual ability, provides for individual learning styles, promotes development through the creative use of their talents, enhances critical thinking skills, and encourages a commitment of service to the community.

Each year, more than a dozen young men from St. Xavier become National Merit Scholars and more than 97 percent of the student population goes on to college after graduation. Brother James M. Kelly, president, doesn't hesitate to give credit elsewhere: "We prepare the soil, plant the seeds, and water them, but it is God who gives them growth."

To give students a leg up in the increasingly competitive college

FROM TOP:
St. Xavier High School was once located at 2nd Street and Broadway.

Brother James M. Kelly, CFX, is president of St. Xavier.

The John L. Hoeck Specialized Studies Center was built in 1984.

environment, St. Xavier provides a learning environment that gives students and teachers experience using the latest technology. St. Xavier is comprised of several buildings spread out on a 28-acre campus. Many of the buildings are new or newly renovated. A special emphasis has been placed on assuring that all students have access to computer labs that are filled with state-of-the-art equipment. The Library and Media Center is filled with reference materials, and also houses an on-line catalog, CD databases, and multimedia workstations.

Individual Attention

Although St. Xavier is a large school, with more than 1,500 students, each student is guaranteed individual attention. "We have been a very large school for more than 50 years, and we make use of several deliberate strategies to break the school down into smaller units," says Sangalli. Some of the attractive features of St. Xavier's structure are small class sizes, a student-to-teacher ratio of 17-to-1, and a team of eight full-time guidance counselors.

One of the advantages of a large school is the number and variety of activities offered. To balance academic demands, extracurricular activities help students develop additional skills. More than 50 clubs and activities have been created in response to the diverse interests and talents of the student body. These activities—including fine performing arts, academic clubs, student council, yearbook, and many others—encourage students to develop all facets of their personalities.

Students are able to participate in virtually any sport, and many teams at St. Xavier are highly competitive. In fact, every year since 1941, the school has won at least one state championship. However, Kelly is quick to point out that "in a school of St. Xavier's caliber, academics always trump athletics, and religious and moral development trump them both."

"St. Xavier students experience a tradition unequaled in academic excellence, in extracurricular programs, and in personalized service," Sangalli says. "And I want every parent to know that we have a place for their son at St. Xavier."

CLOCKWISE FROM TOP LEFT: ST. XAVIER IS COMPRISED OF SEVERAL BUILDINGS SPREAD OUT ON A 28-ACRE CAMPUS.

ST. FRANCIS XAVIER IS PATRON OF THE SCHOOL.

THE SCHOOL'S CHAPEL BECKONS FACULTY AND STUDENTS TO CONNECT DAILY EXPERIENCES TO LIFE'S ULTIMATE QUESTIONS.

The Sullivan Colleges System

Serving students and employers in the Louisville area for more than 135 years, The Sullivan Colleges System continues to be the area's largest and one of its most progressive groups of private career colleges. With more than 6,000 students attending annually, the university and colleges offer a range of programs from short, individual classes to a graduate school offering a masters in business administration. Students have the opportunity to prepare for a variety of careers, ranging from business management to professional chef, from computer engineer to marine mechanic, from nursing to professional nanny.

With a history dating from 1864, The Sullivan Colleges System has been recognized nationally as a leader in postsecondary career education. The system's major schools are Sullivan University, founded in 1864; Spencerian College, founded in Louisville in 1892; and the Louisville Technical Institute, founded in 1961. The Sullivan family has been involved in career education in Louisville since 1926, first with A.O. Sullivan, then with his son, current President A. R. Sullivan, followed by his grandson, Executive Vice President Glenn Sullivan, and his granddaughter Lisa.

A Variety of Educational Options

Originally established as a one-year school of business, Sullivan has undergone many significant changes since its founding, and in 1972 received the authority to offer associate degrees. In 1979, it received its initial accreditation from the Commission on Colleges of the Southern Association of Colleges and Schools. The bachelor's degree programs started in 1990, and a graduate school of business opened in 1997. Sullivan University is currently Kentucky's largest independent four year college or university, with enrollment of more than 3,500 students on campuses in Louisville, Lexington, and Fort Knox, and with teaching locations at community colleges across Kentucky.

In its school of business and office administration, Sullivan offers programs in management, marketing, computer science, legal nurse consultant, professional nanny, paralegal, early childhood management, accounting, office administration, and secretarial specialties. Its National Center for Hospitality Studies division offers degrees in culinary arts, baking and pastry arts, professional catering, hotel/restaurant management, and travel and tourism. In its upper division programs, the school offers bachelor's degrees in business administration, paralegal studies, and hospitality management. In its graduate school, it offers a master in business administration (MBA) degree with concentrations in entrepreneurship, accounting, marketing, management, and dispute resolution. And the college has recently expanded by offering a variety of on-line classes available to students worldwide.

Founded in 1892 by nationally known educator Enos Spencer, Spencerian College has long been

Sullivan University's National Center for Hospitality Studies teaches students through real-life experience. Associate degrees are available in baking and pastry arts, culinary arts, professional catering, travel and tourism, and hotel/restaurant management; a bachelor's degree in hospitality management is also available (top).

Louisville Technical Institute, a Sullivan Colleges System school, offers high-demand careers in computer graphic design, computer engineering technology, computer-aided design drafting (CADD), robotics, marine mechanics technology, and interior design (bottom).

WITH THE VARIETY OF ACCREDITED SHORT-TERM CAREER PROGRAMS AND DEGREES AVAILABLE, THE SULLIVAN COLLEGES SYSTEM SCHOOLS REFLECT AN ONGOING EFFORT TOWARD TEACHING THE SKILLS NEEDED FOR TODAY'S DIVERSE CAREERS IN A CONSTANTLY CHANGING ECONOMY.

recognized as a leading name in business education. In *Beginning of the Business School*, author Charles G. Reigner notes, "The name Spencerian has embedded itself in the consciousness of the American people . . . [I]t is an honored name." Spencerian, now in its 108th year of operation, offers a variety of degree and career programs at its campuses in Louisville and Lexington.

In Louisville, Spencerian offers programs in business office management, computer accounting specialist, clinical assistant, medical assistant, practical nursing, limited medical radiography and others. At its Lexington technical campus, it offers programs in computer-assisted drafting, computer engineering, and computer graphics.

The Louisville Technical Institute has grown from its beginnings in 1961 as an engineering technology and drafting school to become a full-service technical institution with two campus locations in Louisville. It offers associate degree programs in architectural engineering, mechanical engineering and robotics, computer-aided design (CAD), computer graphic design, and computer network administration.

The Interior Design Institute, a division of Louisville Tech, is one of the leading providers of interior designers in Louisville. The marine technology division is a regional leader in the training of technicians qualified to repair both large and small marine engines.

RECEIVING NATIONAL RECOGNITION

Many of these institutions' degree programs have been recognized nationally. For example, students and staff at Sullivan University's National Center for Hospitality Studies have won more than 200 medals and awards in local, regional, national, and international culinary competitions, including recent medals won at the Culinary World Cup in Luxembourg.

On the service side, the National Center for Hospitality Studies operates a full-service, three star gourmet restaurant, Winston's; a professional catering company, Juleps Catering; and a world-class baking facility, The Bakery, all of which serve as training facilities for students and are open to the public to enjoy the quality food and service of Sullivan's outstanding students in this area.

Louisville Tech's Interior Design Institute has designed award winning interiors for Homearamas and other competitions. And the system's Center for Business and Corporate Training is a major resource of customized training, working with numerous Louisville companies to meet their specific training needs, either at one of the college campuses or at the business or industry location.

In 1999, The Sullivan Colleges System was awarded a world-class entrepreneurial company award for service by Ernst & Young and *USA Today*.

With almost 750 faculty and staff members serving its campuses, The Sullivan Colleges System is a major provider of employees for companies not only in Kentucky, but also throughout the world. Focusing on the changing needs of its students, The Sullivan Colleges System will be an educational leader for decades to come.

Torbitt & Castleman

For more than 130 years, Torbitt & Castleman has been a leading provider to the grocery industry. Beginning in Louisville as a small wholesaler, the company has grown into one of the top food manufacturers in the world in the private label, contract packaging, and food service industries. ■ Torbitt & Castleman first began business in Louisville in 1869 as a wholesale grocery operation located on Main Street. Like most commerce of the day, the company utilized the Ohio River to ship and receive goods.

But by the end of the 19th century, Louisville had developed into a thriving transportation center for the railroad industry. Seeing the new possibilities provided by this quick form of transportation, Torbitt & Castleman changed focus, leaving the wholesaling business to become a manufacturer and distributor. The company moved its shop to 10th Street and Magnolia Avenue to provide more room, and began producing and packaging syrups for sale to retail stores and wholesalers.

Growth of a Product Line

Today, syrup still plays an important part in the Torbitt & Castleman line of products, and although many of its products are now shipped via trucks, the railroad is still a vital component in the company's continued success. Yet, Torbitt & Castleman realized in the 1970s that to continue to grow, it would have to provide more than just syrup. In order to do that, the firm needed more space.

In 1975, Torbitt & Castleman found the perfect location, purchasing 150 acres of land in Oldham County near Interstate 71. There, the company had unlimited room for growth, strategically located close to a main expressway and an existing railroad line.

With the new location, Torbitt & Castleman wasted little time in expanding its product line. The company that had previously only manufactured syrup began to manufacture a full range of other sauces and condiments. Today, Torbitt & Castleman is a leading manufacturer in the retail private label industry. Although the company's products are on store shelves around the country, consumers will never find a Torbitt & Castleman brand product. Instead, the firm's products are sold under grocers' private labels, and are matches for many national brand products.

The company's complete product line includes a full range of wet goods such as pancake and waffle

Torbitt & Castleman has modern, flexible, high-speed facilities that are used to service the private label food industry.

syrups, chocolate sauces and other flavored dessert sauces, jellies/preserves and fruit spreads, barbecue sauces, Mexican sauces, and steak sauces. Torbitt & Castleman also produces dry mixes such as sweetened instant tea mix, breakfast drink mix, powdered fruit drink, flavored drink mix for milk, gelatin and pudding mix, and pancake mix. Nutritional bars and other nutraceutical products were added to the firm's product line in the late 1990s.

State-of-the-Art Capabilities

Torbitt & Castleman's production site in Oldham County provides more than 300,000 square feet of state-of-the-art manufacturing facilities. The plant is American Institute of Baking (AIB) certified and earned AIB's superior rating in 1999. Approximately 250 employees at the plant work to ensure that every product meets rigorous quality requirements.

The multiple production lines at Torbitt & Castleman's facility use state-of-the-art technology to blend, cook, and package the company's products. On any given day, the plant produces and bottles a multitude of sauces, syrups, and jellies. Some product lines are even interchangeable, capable of producing barbecue sauce, then—after a thorough cleaning—chocolate syrup. Each product line has rigorous standards to ensure superior quality.

Torbitt & Castleman was one of the first manufacturers in the United States to introduce plastic syrup bottles. The firm was also one of the first to introduce plastic packaging for drink mix containers. Plastic is more convenient for consumers and safer for the environment, utilizing a high content of recycled materials that can be recycled again.

The expert research and development team at Torbitt & Castleman is another key to the company's success. The firm's in-house team of research and development specialists has more than 50 years of combined experience. Recognizing the special needs of the private label industry, the team works to provide products that are matches for name brand products. Additionally, the team works to create unique formulations that some customers require, such as recently developed fruit syrups and sugar-free pancake syrup, as well as mustard-based barbecue sauces.

In addition to a high-quality product line, Torbitt & Castleman provides superior distribution, thanks in part to its centralized location in the Greater Louisville area. The company is able to provide a quick turnaround to clients on a national basis, typically providing goods within 10 days of receipt of order.

This privately owned food manufacturer understands the importance of being a good corporate citizen. Torbitt & Castleman provides access road and water and sewer services for the New Oldham County YMCA, and supports Oldham High School and various other activities in the community.

Torbitt & Castleman has been a vital part of the Greater Louisville community and a forward-thinking, quality provider for the grocery industry for more than 130 years. As the grocery industry continues to change and require new products, Torbitt & Castleman will continue to meet its needs.

Clockwise from top left:
The plant is American Institute of Baking (AIB) certified and earned AIB's Superior rating in 1999.

Quality and Assurance performs routine product checks to see that standards are met to ensure superior and consistent quality.

Conveniently located in the heart of the country, Torbitt & Castleman is able to provide quick delivery to customers.

The Kroger Company

THE KROGER COMPANY IS THE LARGEST FOOD RETAILER IN America, with more than 2,200 food stores in some states under the Kroger name and in other states under names such as Ralphs, Fred Meyer, Smith's, Fry's, King Sooper's, City Market, and many others. A total of 1,700 convenience stores are also part of Kroger's holdings, and the company now owns the nation's largest jewelry chain, consisting of more than 300 stores acquired from Fred Meyer.

In the Louisville area, the company operates more than 30 state-of-the-art combination food stores. Customers can always count on Kroger for the most innovative concepts in food retailing, and the company is dedicated to remaining an important part of local neighborhoods so that customers don't have to go out of their way to shop.

"It's a different world today in the food business," says John Hackett, president of Kroger's Louisville marketing area. "You have to stay on top of what the customer is looking for at all times and offer it at a competitive price. And in today's world, our customers are looking for convenience and shortcuts. This is especially true since such a large number of women joined the workforce. The other part of the equation is being an important part of the communities that you serve—their causes, their problems, and their activities."

A Tradition of Involvement in Louisville

Carrying its slogan Count On Us beyond the doors of its stores, Kroger takes part in a variety of community and charitable activities every year. Kroger is proud of its heritage of being a civic leader, maintaining involvement in many different activities. The company serves as the number one financial supporter of the Kentucky Derby Festival and has created the Kentucky Derby Garland of Roses since 1987. Kroger is also the original creator of the Lilies for the Fillies garland for the winner of the Kentucky Oaks, and the company creates the seven garlands needed for the Breeders' Cup winners annually.

Kroger's involvement in Louisville goes far beyond cash registers, produce, and shelves of groceries; Kroger goes to the heart of the community. The company serves as a major contributor to the Dare to Care Food Bank, Crusade for Children, United Way, Muscular Dystrophy Association, Operation Brightside, University of Louisville,

THE KROGER COMPANY'S PENGUIN BALLOON PARTICIPATES IN THE DERBY FESTIVAL GREAT BALLOON RACE.

University of Kentucky, Bellarmine College, Salvation Army Angel Tree, Louisville Zoo, Louisville Riverbats, Greater Louisville Inc., Boy Scouts and Girl Scouts, Louisville Urban League, and Bridgehaven. Kroger also supports other community activities such as the Boys Club and Girls Club, Jefferson County Public Education, Fund for the Arts, Kentucky Center for the Arts, Better Business Bureau, Jefferson County Historical Society, and Leadership Louisville.

Leading Supermarkets into the Next Century

Kroger has always been a company to lead the way in store design, variety, service, and price in the market. In keeping with that approach, Kroger is introducing the Century Market, a unique design, which will be extended throughout the United States.

Century Market is more than just a grocery store. It is a store within a store, designed to better serve its customers and provide them with true one-stop shopping. Consumers will find a huge variety of unique shops and departments, as well as conveniences such as state-of-the-art pharmacies with drive-through service, gas stations, and a florist shop that will also deliver.

Kroger offers several different shops that cater to customers looking for a ready-to-eat meal. Customers on the go can serve themselves from salad bars that include soups and pasta salads, or purchase delicious delicacies from fresh pastry departments, New York-style delis, and Chef Shoppes offering prepared gourmet foods. Other specialty sections provide cheeses, made-to-order pizza, European crusty breads and hot breads, and farmer's market produce. Although the traditional butcher shops and fresh seafood sections have been a part of Kroger for some time, today they offer far more variety and service than ever before.

But Kroger's Century Market isn't limited to food items. The stores also offer nutrition centers, card and party shops, bath and beauty products, health and general merchandise, books and magazines, baby care centers, expanded pet centers, and customer service centers. Many Kroger stores even offer financial centers.

Set to serve Kroger customers' shopping needs well into the future, Kroger Century Market's many one-stop shops are sure to keep the company on the leading edge of its industry.

EACH YEAR, KROGER DESIGNS THE GARLAND OF ROSES FOR THE KENTUCKY DERBY.

Norton Healthcare

Since 1886, the name Norton has been synonymous with excellence in heath care, and over the past century, Norton Healthcare has expanded and evolved to meet the needs of a growing region. Today, the system includes six owned hospitals, 17 leased or managed hospitals, five immediate care centers, 8,000 employees, and nearly 2,000 active members of the medical staff, providing an unparalleled network of services to Kentucky and southern Indiana.

Norton Healthcare traces its start to 1886 when the John H. Norton Memorial Infirmary opened its doors in Louisville on the corner of Third and Oak streets, where it remained for its first 88 years. As Louisville's first self-supported hospital, the Norton infirmary provided outstanding care by maintaining a high-quality staff, focused on providing the best possible patient care. The medical staff included one of the early presidents of the American Medical Society, Louis Samuel McMurtry, and the hospital housed one of the nation's first schools of nursing.

The Development of a Health Care Network

The first step toward expansion began in 1969, when Norton merged with the financially struggling Children's Hospital. Children's Free Hospital, as it was originally called, opened in 1892 on East Chestnut Street as a hospital devoted to children, rich or poor, with any illness or injury. Immediately, the hospital became the key pediatric treatment center in an impoverished region, building a reputation for care that it carries with it today. In 1973, both hospitals moved into a new building on East Chestnut Street in the Louisville Medical Center.

Then, in 1981, Children's Hospital and Kosair Crippled Children Hospital, established in 1926 by Kosair Shrine Temple, joined to form Kosair Children's Hospital. At this time, Norton-Children's Hospitals Inc. became known as NKC Inc. Kosair Crippled Children Hospital on Eastern Parkway had become well known for its treatment of children with orthopedic disorders and, during the late 1930s and early 1940s, played an important role by treating children who were victims of the polio epidemic. In 1986, Kosair Children's moved into a new building on East Chestnut Street, forming the region's only full-service, freestanding children's hospital. Today, Louisville often receives worldwide attention via the reputation of Kosair Children's, heralded as one of the nation's 10 best hospitals for children by *Child* magazine in the magazine's only such evaluation.

In 1989, the organization grew again when NKC Inc. merged with Methodist Evangelical Hospital on East Broadway, forming Alliant Health System. Methodist Evangelical Hospital, which opened in 1960, had developed a reputation as a modern, high-tech hospital. Today, the facility is known as the Norton Healthcare Pavilion and is dedicated primarily to outpatient

CLOCKWISE FROM TOP:
NORTON HEALTHCARE TRACES ITS START TO 1886 WHEN THE JOHN H. NORTON MEMORIAL INFIRMARY OPENED ITS DOORS IN LOUISVILLE ON THE CORNER OF THIRD AND OAK STREETS, WHERE IT REMAINED UNTIL 1973. NORTON HOSPITAL IS NOW ON CHESTNUT STREET IN THE LOUISVILLE MEDICAL CENTER.

RECOGNIZED AS A REGIONAL LEADER IN THE TREATMENT OF CANCER AND HEART DISEASE, NORTON HEALTHCARE ALSO DELIVERS MORE BABIES THAN ANY OTHER HEALTH CARE ORGANIZATION IN LOUISVILLE.

FOR PATIENTS WITH SPINAL INJURIES AND DISEASES, THE NORTON HOSPITAL LEATHERMAN SPINE CENTER FOCUSES ON A MULTIDISCIPLINARY APPROACH TO DETERMINE THE MOST EFFECTIVE TREATMENT PROGRAM, INCLUDING MEDICAL AND SURGICAL OPTIONS. THE SPINE CENTER WAS THE SITE OF AMERICA'S FIRST USE OF THE COTREL-DUBOUSSET SURGICAL PROCEDURE FOR SCOLIOSIS REPAIR, WHICH IS NOW STANDARD PROCEDURE THROUGHOUT THE COUNTRY.

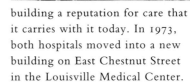

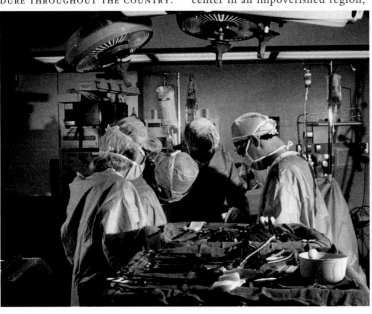

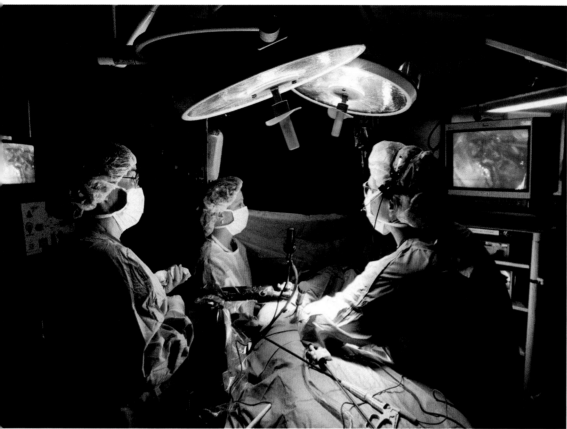

CLOCKWISE FROM TOP: SURGEONS PERFORM A MINIMALLY INVASIVE LAPAROSCOPIC PROCEDURE AT NORTON HOSPITAL. NORTON HEALTHCARE IS A PARTNER WITH THE UNIVERSITY OF LOUISVILLE SCHOOL OF MEDICINE IN THE CENTER FOR ADVANCED SURGICAL TECHNOLOGIES, WHICH IS DEDICATED TO THE DEVELOPMENT AND TEACHING OF MINIMALLY INVASIVE TECHNOLOGIES FOR THE TREATMENT AND MANAGEMENT OF DIGESTIVE, VASCULAR, BREAST, PULMONARY, AND OTHER DISEASES.

THE EMERGENCY DEPARTMENT AT NORTON AUDUBON HOSPITAL ON POPLAR LEVEL ROAD IS ONE OF THE STATE'S BUSIEST. THE DEPARTMENT WAS RECENTLY EXPANDED, TAKING ADVANTAGE OF THE LATEST AVAILABLE TECHNOLOGY. ALL PATIENTS ARE TREATED IN PRIVATE ROOMS.

THE WOUND CARE CENTER AT NORTON SUBURBAN IN ST. MATTHEWS PROVIDES HELP AND HOPE FOR CHRONIC, NONHEALING WOUNDS USING THE MOST ADVANCED MEDICAL RESEARCH AND TECHNOLOGY.

care. The pavilion houses centers for excellence for the treatment of diabetes, cancer, and orthopedic disorders. Norton Healthcare Pavilion also has a surgical outpatient center, as well as Hospice and Palliative Care of Louisville's only inpatient unit.

REACHING OUT TO THE COMMUNITY

Throughout much of the 20th century, Norton focused its energies on its network of hospitals, all centered in the downtown Louisville Medical Center. But in 1998, four new hospitals joined the network, expanding the organization's presence into the suburbs and beyond. The hospitals—Norton Audubon, Norton Suburban, Norton Southwest, and Norton Spring View (located in Lebanon, Kentucky)—all had rich histories of community service and strong reputations of their own.

Norton Audubon Hospital has roots in one of the oldest organized health care facilities in Louisville. In 1832, what today is Norton Audubon began as a clinic to combat a cholera epidemic. It grew into St. Joseph's Infirmary on Eastern Parkway, owned and operated by the Sisters of Charity of Nazareth. In 1970, the hospital was purchased by Humana Inc. When it moved 10 years later to its current facility on Poplar Level Road, it was known as Audubon Hospital. The hospital achieved worldwide recognition for its heart institute, where an experimental mechanical heart implant was developed and used, and for its sleep disorders center, which is one of the oldest accredited centers of its kind in the country.

Norton Suburban opened its doors in 1972 and was the first hospital in eastern Jefferson County. Norton Suburban opened the state's first freestanding outpatient surgery center in 1975 and acquired Kentucky's first lithotripter in 1985. Today, the hospital has one of the

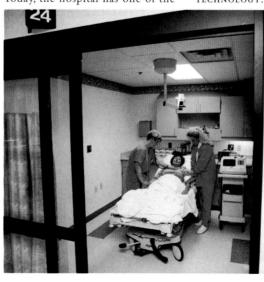

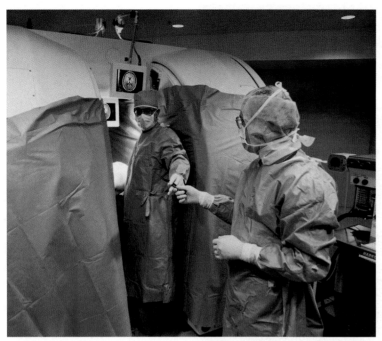

Using an intraoperative MRI center—the ultimate in computer-assisted, image-guided surgery—physicians from Norton Hospital and Kosair Children's Hospital view MRI images live and in real time during surgery. This capability allows surgeons to be more precise, shortens recovery time, improves surgical results, and lowers medical costs.

Dedicated to the treatment of ill and injured children, Kosair Children's Hospital includes specialties in pediatric anesthesiology and pediatric surgery, including cardiac surgery, neurosurgery, orthopedic surgery, and a kidney and heart transplant program.

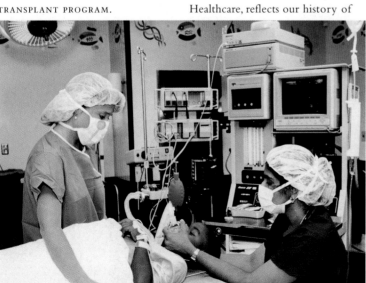

region's busiest urodynamic labs, a maternal/fetal diagnostic lab, a thriving obstetrics program, a Level III neonatal intensive care unit (NICU), and the area's only wound care center.

Norton Southwest and Norton SpringView also grew out of a commitment to bringing quality hospital care to their respective communities, and both facilities offer emergency services, surgery, and a skilled nursing unit, and are known for their dedicated service to suburban and rural residents.

After the purchase of the four hospitals, Alliant Health System changed its name to better reflect its history. In January 1999, Alliant Health System became Norton Healthcare. "Our community strongly identifies with the Norton name, and we believe our name, Norton Healthcare, reflects our history of delivering quality patient care to the people in our community," says Stephen A. Williams, Norton Healthcare president and CEO.

For more than a century, the Louisville community has benefited from the combined experience of the hospitals and other medical facilities that make up Norton Healthcare. Patients benefit from the advanced centers of specialized care, and the organization's regional presence allows easy access to a Norton hospital for all residents. Drawn by the innovative and high-tech approaches used by many Norton services, patients have come to Norton Healthcare from across the nation for treatment of heart disease, cancer, and other ailments.

Prevention and Care through Specialty Services

As part of a dedication to serving the needs of the community, Norton Healthcare offers specialized services that focus not only on treatment, but also on prevention and support. For example, the Norton Healthcare Cancer Resource Center offers the latest information on the prevention and detection of cancer, treatment options, and recovery. Specially trained oncology nurses on staff at the center provide personal service to patients and family members looking for answers to their questions and comforts for their fears. The center continually has been recognized as one of the best and most beneficial services of its kind in the country.

Another specialty area that has brought attention to Norton Healthcare is its state-of-the-art heart care. Norton, Kosair Children's, and Norton Audubon hospitals have built outstanding reputations for their treatment of adult and pediatric heart patients. Surgeons are pioneering work in the use of laser surgery to treat heart disease and in minimally invasive treatments of heart defects in children and adults.

Specialized Surgery and Treatment Centers

All Norton hospitals are proud to employ some of the most advanced surgical techniques available, and, as part of a commitment to technology, Norton Healthcare is a proud partner with the University of Louisville School of Medicine in the Center for Advanced Surgical Technologies. This program is dedicated to the study and development of minimally invasive technologies for the treatment and management of digestive, vascular, breast, pulmonary, and other diseases. In January 2000, it became one of only four centers in the United States to house Storz laparoscopic teaching stations, which allow surgeons and residents to refine surgical skills in a virtual setting.

Norton Hospital also is one of only 11 centers worldwide and four in the United States with an intraoperative magnetic resonance imaging (IMRI) center. The ultimate in computer-assisted, image-guided surgery, the IMRI enables physicians to view MRI images live and in real time during surgery. This allows physicians to be more precise during surgery, shortens recovery time, improves surgical results, and lowers medical costs.

Also part of Norton Healthcare is the Leatherman Spine Center, part of the Norton Hospital Spine and Neuroscience Center. The Leatherman Spine Center uses a multidisciplinary approach to evaluate a patient's spine problem to determine the most appropriate and effective level of care. The center offers a full range of treatments from pain management to surgery, as well as scoliosis screenings.

Although Norton Healthcare is well known for its many specialized treatment programs, the company is also proud of its outstanding inpatient and emergency room services. At all Norton Healthcare hospitals, a high priority is placed on patients who receive focused care from a high-quality, patient-centered staff. Emergency department services at each hospital always rank among the top priorities of care.

Quality Services for Women and Children

Norton Healthcare was one of the first such organizations in the nation to establish a center exclusively for women's health. The Norton

Hospital Women's Pavilion provides a wide range of care, including routine and high-risk maternity services; women's surgery; and breast, continence, health and resource, and heart outpatient centers. The Norton Hospital Women's Pavilion is dedicated to the belief that women should take active responsibility for their own health, and a major component in making that possible is providing women with the facts they need to make informed health decisions. As part of the pavilion's services, the Norton Suburban Birthing Center offers a convenient east-end location focused on women's care, including a dedication to all aspects of maternity from education to regular and high-risk deliveries.

The national spotlight often falls on Norton Healthcare in terms of specialized care for children. Kosair Children's Hospital has long been regarded as an exceptional facility where children, adolescents, and young adults receive highly advanced care for everything from cuts and fevers to life-threatening illnesses. The facility is a national leader in offering advanced procedures, techniques, and equipment in pediatrics, and was designated as one of the nation's 10 best hospitals.

Kosair Children's boasts of one of the nation's largest intensive care nurseries and has the region's only pediatric trauma center. The hospital treats all children who need its care, regardless of a family's ability to pay. Kosair Children's is the pediatric teaching facility for the University of Louisville School of Medicine, and includes special units for cancer, orthopedics, diabetes, psychiatry, and pediatric surgery, in addition to its highly specialized pediatric intensive care unit and its infectious disease and other programs.

"Many people don't realize that the treatment of children is not just smaller-scale treatment of an adult. It is completely different health care," says Williams. "Children need closer monitoring and they can't always tell a doctor or a nurse where it hurts. Kosair Children's is completely geared for children, from the many different small-sized medical instruments we use to the colorful paintings surrounding the hospital and the sky scenes that we paint on the ceilings."

MORE THAN JUST HOSPITALS

Part of Norton Healthcare's goals through the years has been to bring quality services closer to everyone," Williams says. "We approach that goal in a multifaceted way."

Many in the community are served by the Norton Immediate Care Centers. There are five immediate care centers throughout Jefferson County that provide care for patients who don't have a primary care physician or who need after-hours care for an injury that doesn't require hospital care. Norton Healthcare offers many other services that directly benefit the community, including helping people find a qualified physician through the physician finder referral service, offering medical education seminars open to the public, and managing practices for more than 30 physician groups. Norton Healthcare is part of University Medical Center, which manages the University of Louisville Hospital.

Since 1886, Norton has been dedicated to providing the best possible care for its community. As the 21st century brings new challenges and new technologies, Norton will continue to position itself on the leading edge of health care. The organization's commitment to local residents will continue to shape the future of medical care in the region and will continue to define the many ways in which Norton Healthcare meets the needs of the Louisville area.

FIVE IMMEDIATE CARE CENTERS HELP MAKE UP THE NORTON HEALTHCARE NETWORK. THE CENTERS PROVIDE TREATMENT FOR ILLNESSES AND MINOR INJURIES AND OFFER AFTER-HOURS CARE (TOP).

THE NEONATAL INTENSIVE CARE UNIT (NICU) AT KOSAIR CHILDREN'S HOSPITAL PROVIDES EXPERT CARE FOR THE TINIEST NEWBORNS AND THOSE WITH SPECIAL NEEDS. THE UNIT IS CONNECTED BY PEDWAY TO THE NORTON HOSPITAL WOMEN'S PAVILION, ENSURING AT-RISK NEWBORNS CAN BE TRANSPORTED QUICKLY TO THE NICU (BOTTOM).

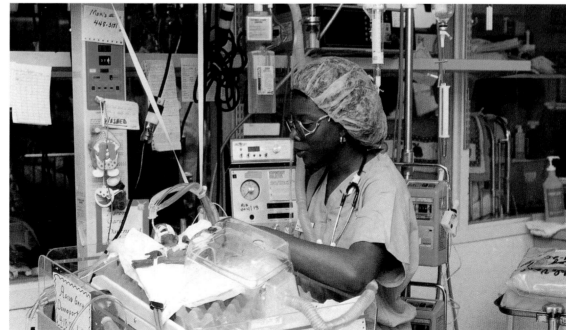

Kentucky Fair and Exposition Center/Kentucky International Convention Center

In 1950, the Kentucky State Fair Board, an agency of the Kentucky Tourism Development Cabinet, began construction on one of the greatest and largest exposition facilities of its kind in the world—the Kentucky Fair and Exposition Center (KFEC). Today, the Kentucky State Fair Board is still at the forefront of Kentucky tourism development, operating both KFEC and the Kentucky International Convention Center, two of the largest multipurpose facilities in the state. Together, these two facilities provide more than $400 million of revenue annually through conventions, trade shows, and other events.

Kentucky Fair and Exposition Center

KFEC opened its doors for business in 1956 as the largest multipurpose facility in the world, hosting the 1956 Kentucky State Fair. Appropriately, it was the state's agriculture industry that pushed for the creation of such a facility.

"This facility is truly a product of the Kentucky agriculture and agribusiness industry," says Harold Workman, president and CEO of the Kentucky State Fair Board. "It was built as a facility to showcase agriculture in Kentucky, and that continues to be a major part of our business."

Some of the country's largest agribusiness-related shows call KFEC home, including the National Farm Machinery Show, North American International Livestock Exposition, and World's Championship Horse Show, as well as numerous other livestock events. In addition, the Future Farmers of America (FFA), an organization dedicated to providing agricultural education to the youth of America, is the largest convention hosted in Louisville, generating more than $20 million in economic impact.

While agriculture helped create KFEC, its design as a multipurpose building allows it to rank as one of the most-used facilities of its type. Boasting more than 1 million square feet of indoor exhibit space and more than 2 million square feet of outdoor exhibition space, the center is one of the 10 largest convention facilities in the United States, competing on the same level as larger cities such as Chicago, Atlanta, and Las Vegas.

On any given day, the space may be filled with a wide range of events, from trade shows and conventions to livestock shows, concerts, and community events. The facility also contains Freedom Hall, which is home to sporting events such as University of Louisville basketball games and Louisville Panthers ice hockey. KFEC's staff is renowned for its ability to turn an auditorium filled with thousands of pounds of dirt for a horse show into an auditorium prepared for a concert—overnight.

Kentucky International Convention Center

The dawn of the 21st century was celebrated in Louisville in a newly renovated downtown convention facility. After undergoing three years of work, the transformed convention center doubled in size—now containing 200,000 square feet of exhibit space, more than 50 meeting rooms, and a 30,000-square-foot ballroom.

The transformation was so complete that the facility—once known as the Commonwealth Convention Center—is today the Kentucky International Convention Center. "From the supertruss towering over the two city blocks it spans, down to the brickwork and wall of win-

On any given day, the Kentucky Fair and Exposition Center may be filled with a wide range of events, from trade shows and conventions to livestock shows, concerts, and community events. The facility also contains Freedom Hall, which is home to sporting events such as University of Louisville basketball games and Louisville Panthers ice hockey.

dows overlooking tree-lined sidewalks, the Kentucky International Convention Center has definitely taken on a new identity," says William M. Kuegel Sr., chairman of the Kentucky State Fair Board.

As a result of its transformation, this modern facility can compete for more international and national conventions, while offering the benefits of being in a convenient downtown location. The Kentucky International Convention Center already hosts international conventions for several major organizations, including Papa Johns International Inc., International Porcelain Artists, Starlight International, and the World Aquaculture Society.

The new convention center provides far more than just additional convention space. Its awe-inspiring architecture strives to showcase Louisville's heritage. The downtown skyline is accented by the building's supertruss, which allows the Kentucky International Convention Center to span a city street covering two urban blocks. The supertruss resembles the bridges crossing the Ohio River a few blocks away.

Once inside the building, the river imagery continues on the blue-gray terrazzo floor that winds through the lobbies. Along the edges of the terrazzo floor are pictorial illustrations of many of Louisville's landmarks. The formal ballroom provides excellent facilities for upscale corporate and community events. Overall, the elegance of the facility sets it apart from the competition.

The Kentucky International Convention Center provides an edge for the city to attract more convention and trade shows than ever before. And while the facility is indeed beautiful, event planners around the world love its functional design. The functional aspects of the facility include the ability for semi-trucks to actually drive into the building, allowing event supplies to be easily unloaded inside the exhibition areas.

Combined, the Kentucky Fair and Exposition Center and the Kentucky International Convention Center, under the leadership of the Kentucky State Fair Board, are poised to lead Louisville well into the next century as two of the best multipurpose facilities in the United States.

AFTER UNDERGOING THREE YEARS OF WORK, THE KENTUCKY INTERNATIONAL CONVENTION CENTER NOW CONTAINS 200,000 SQUARE FEET OF EXHIBIT SPACE, MORE THAN 50 MEETING ROOMS, AND A 30,000-SQUARE-FOOT BALLROOM. THE DOWNTOWN SKYLINE IS ACCENTED BY THE BUILDING'S SUPERTRUSS, WHICH RESEMBLES THE BRIDGES CROSSING THE OHIO RIVER A FEW BLOCKS AWAY. ONCE INSIDE THE BUILDING, THE RIVER IMAGERY CONTINUES ON THE BLUE-GRAY TERRAZZO FLOOR THAT WINDS THROUGH THE LOBBIES. ALONG THE EDGES OF THE TERRAZZO FLOOR ARE PICTORIAL ILLUSTRATIONS OF MANY OF LOUISVILLE'S LANDMARKS.

Jewish Hospital HealthCare Services

In 1903, a group of Jewish civic leaders founded Jewish Hospital with a mission to treat all patients with the highest-quality care and to provide opportunities for medical research and education. Since opening its doors in 1905, the hospital has continually reaffirmed that commitment, and as the needs of the communities it serves have evolved, so has Jewish Hospital's resolve and ability to meet them.

From a 32-bed hospital located at Floyd and Kentucky streets to a diversified health care company, Jewish Hospital HealthCare Services (JHHS) is on the leading edge of medicine. The health care provider's symbol, the Circle of Care, represents JHHS as a regional network with 7,400 full-time employees at more than 35 health care locations throughout Kentucky and southern Indiana. With 1,800 beds serving more than 1.4 million patients annually, network facilities deliver a continuity of care that allows consumers easy access to premier services in every specialty while they remain close to home.

Taking Health Care to the People

JHHS began building its regional health care network in the early 1980s. Four Courts Senior Center, a long-term-care facility founded as the Louisville Hebrew Home, was the first organization to join the network in 1983. In 1988, JHHS began managing the Kings' Daughters Hospital, which was purchased in 1992 and renamed Jewish Hospital Shelbyville. Indiana's Scott Memorial Hospital in Scottsburg and Washington County Memorial Hospital in Salem joined JHHS in 1991; Clark Memorial Hospital came on board the following year. In Kentucky, Pattie A. Clay Regional Medical Center in Richmond joined in 1993, and the Frazier Rehab Institute merged with JHHS in 1994. Southern Indiana Rehab Hospital has been managed by Frazier Rehab Institute and owned in partnership with Floyd Memorial Hospital and Clark Memorial Hospitals since 1994. Taylor County Hospital at Campbellsville, Kentucky, became part of the system in 1995, and the Visiting Nurse Association joined in 1996. JHHS has comanaged the University of Louisville Hospital since 1996.

By expanding its range of high-tech medical services, enhancing facilities, and improving financial performance, JHHS has enabled all of the institutions in its network to deliver a full range of the highest-quality medical services to people who need them—in their own communities.

Centers of Excellence

As the leader in medical care in this region for nearly 100 years, Jewish Hospital is internationally known. The flagship of the JHHS network, Jewish Hospital is a 442-bed, high-tech, tertiary-care regional referral center with unique expertise distinguished by leading-edge advancements in 10 medical and surgical specialties or Centers of Excellence. Attracting patients from around the world, these special services include hand and microsurgery, organ transplantation,

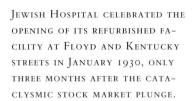

Jewish Hospital celebrated the opening of its refurbished facility at Floyd and Kentucky streets in January 1930, only three months after the cataclysmic stock market plunge.

The future of health care in Louisville lies in the city's medical center. Jewish Hospital HealthCare Services (JHHS) has made medical education and research a number one priority.

JOSEPH E. KUTZ, M.D. (LEFT) AND HAROLD E. KLEINERT, M.D. FOUNDED KLEINERT, KUTZ AND ASSOCIATES HAND CARE CENTER PLLC (KKA), INTERNATIONALLY RENOWNED FOR ITS INNOVATIONS AND PROCEDURES IN REPLANTATION AND REPAIR OF THE UPPER EXTREMITIES. WELL OVER 900 KLEINERT FELLOWS FROM 49 DIFFERENT COUNTRIES HAVE TRAINED WITH THE HAND CARE GROUP SINCE 1960.

THE NATION'S FIRST SUCCESSFUL HAND TRANSPLANT WAS PERFORMED BY PHYSICIANS FROM KKA AND THE UNIVERSITY OF LOUISVILLE AT JEWISH HOSPITAL ON JANUARY 25, 1999. WARREN C. BREIDENBACH, M.D. (LEFT), A PARTNER IN KKA AND ASSISTANT CLINICAL PROFESSOR OF PLASTIC AND RECONSTRUCTIVE SURGERY AT THE UNIVERSITY OF LOUISVILLE, AND JON W. JONES JR., M.D. (RIGHT), ASSISTANT PROFESSOR IN THE DEPARTMENT OF SURGERY AT THE UNIVERSITY OF LOUISVILLE AND THE PANCREAS TRANSPLANT PROGRAM DIRECTOR FOR THE UNIVERSITY OF LOUISVILLE AND JEWISH HOSPITAL, SHARE THE DAIS WITH HAND TRANSPLANT RECIPIENT MATTHEW SCOTT (CENTER) AT A PRESS CONFERENCE MARKING THE ONE-YEAR ANNIVERSARY OF HIS LANDMARK SURGERY.

primary care, heart and lung care, rehab medicine, neuroscience, outpatient care, home health, occupational health, and plastic and aesthetic surgery.

The Transplant Center began in the 1950s as a collaborative effort of Jewish Hospital and the University of Louisville School of Medicine. It is one of an elite group of centers performing all five solid organ transplants, and was 12th in the nation to be designated a Medicare-certified liver, lung, heart/lung, and kidney transplant center.

The Transplant Center continues to achieve medical firsts. The first kidney transplant in western Kentucky was performed there in 1964, as well as the nation's first use of a Thoratec ventricular heart assist device (VAD) in 1985. The state's first heart, pancreas, heart/lung, adult liver, kidney/liver, double lung, and laparoscopic donor nephrectomy transplants also took place at Jewish Hospital.

Hand and Microsurgery became Jewish Hospital's first Center of Excellence in 1980, and today, Kleinert, Kutz and Associates Hand Care Center PLLC (KKA) is internationally renowned for its innovative procedures in replantation and repair of the upper extremities. Achievements at KKA include the first reported repair of a digital artery, the first bilateral upper arm and forearm replantation, one of the world's first cross-hand replantations, and pioneering work in reconstruction using free tissue transfer.

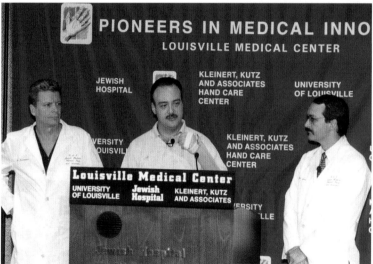

The first hand transplant in the United States was performed by KKA and University of Louisville surgeons at Jewish Hospital on January 25, 1999. Just three months later, transplant recipient Matthew Scott was able to throw out the first pitch at his beloved Philadelphia Phillies' season-opening game.

JEWISH HOSPITAL: A LEADER IN HEART CARE

Jewish Hospital's Heart and Lung Institute, essentially a hospital within a hospital, opened in 1984, and provides the highest-quality patient care, research, and education. In 1995, the institute moved into the new, state-of-the-art Rudd Heart and Lung Center, a 14-story symbol of the hospital's rank as one of the largest cardiac surgery centers in the nation.

Pioneering work in this field began with the first adult open-heart surgery in western Kentucky, performed at Jewish Hospital in 1965. Institute achievements include establishing Kentucky's first hyperbaric oxygen treatment facility and performing the first angioplasty, adult heart transplant, ventricular remodeling, port-access mitral valve surgery, liquid ventilation procedure, use of the Bird high-frequency ventilator, and minimally invasive saphenous vein harvest.

In August 1999, a surgical team led by University of Louisville/Jewish Hospital physicians conducted final preclinical trials for implanting ABIOMED Inc.'s AbioCor in patients at risk of death from an irreparably damaged heart.

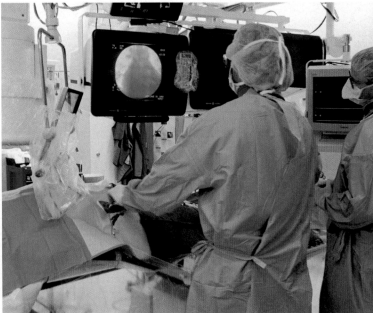

CLOCKWISE FROM TOP LEFT: LAMAN A. GRAY JR., M.D., A UNIVERSITY OF LOUISVILLE AND JEWISH HOSPITAL CARDIAC SURGEON, PERFORMS MINIMALLY INVASIVE CARDIAC SURGERY.

AN ANGIOPLASTY PROCEDURE IS CONDUCTED IN THE STATE-OF-THE-ART CARDIAC CATHETERIZATION LABORATORY AT JEWISH HOSPITAL.

HEALTH CARE EXPRESS' CUSTOM-DESIGNED VAN CAN BE SEEN ON THE SIDELINES AT SPORTING EVENTS AROUND TOWN—ANOTHER VISIBLE SIGN OF JEWISH HOSPITAL AND FRAZIER REHAB INSTITUTE'S COMMITMENT TO SERVE THE COMMUNITY'S HEALTH CARE NEEDS.

Jewish Hospital's selection as one of ABIOMED's test sites confirms its leadership in heart care, technology, and transplantation.

ENHANCING SERVICES FOR EVER EXPANDING NEEDS

Committed to premier care, Jewish Hospital opened the country's first totally dedicated Plastic and Aesthetic Surgery Center and Kentucky's first freestanding ambulatory surgery center in 1986. During the 1990s, the hospital established Kentucky's first Emergency Heart Center and first Emergency Stroke Center, as well as the first full-service hotel and recovery center affiliated with a hospital, The Inn at Jewish Hospital. The organization continues to provide a diversity of services that enhance care for patients and meet their ever expanding needs.

One such service provider is Frazier Rehab Institute, a nationally recognized rehab hospital, which sets the standard of excellence in physical rehabilitation services. The interdisciplinary team at Frazier's expansive network of inpatient and outpatient facilities in Kentucky and southern Indiana helps people overcome their physical disabilities.

Through its partnership with Jewish Hospital and Frazier Rehab Institute, the University of Louisville's comprehensive sports medicine program has become a national model of care and innovation. This unique collaboration provides training, injury prevention, and immediate, comprehensive medical treatment for all student athletes. Certified athletic trainers also provide on-site coverage for Kentuckiana high school athletes. In 1999, Jewish Hospital and Frazier Rehab Institute introduced Health Care Express™, a custom-designed, 40-foot van outfitted with high-tech equipment and staffed by medical personnel able to provide on-site health care to athletes and fans alike.

In addition, the Visiting Nurse Association (VNA) has brought compassionate care into people's homes 24 hours a day, seven days a week. The services VNA provides combine well with Jewish Hospital's acute care, occupational medicine, rehabilitation, and home medical equipment services to create a comprehensive continuum of care for people in the region.

Jewish Hospital also provides five-star hotel services and premier medical care at the Trager Pavilion. Opened in October 1999, Trager's comfortable patient suites provide a quiet, secluded setting; direct admission to the floor; concierge services; gourmet dining services provided by a noted, full-time chef; mid-morning and afternoon gourmet break; daily newspapers; and more.

Another aspect of Jewish Hospital's comprehensive care is SKYCARE, Kentucky's first

hospital-based ambulance service. When it was launched in 1982, the helicopter became a symbol of the hospital's entry into the world of high tech. In 1997, to keep pace with rapidly changing needs, Jewish Hospital and other members of the Louisville Medical Center consolidated their existing air ambulance programs to form StatCare, a nonprofit corporation to provide the region with quality emergency services. In 1999, StatCare made more than 1,000 trips, and more than 79 percent of those patients were from outside the metropolitan Louisville area.

Recognizing the value of education in preventive health care, Jewish Hospital opened the state's first Healthy Lifestyle Center at Bashford Manor Mall in 1989 to provide shoppers with medical information. Centers at Oxmoor and Jefferson malls soon followed. The fourth and newest center opened in 1997 at the Jewish Community Center. Since the program's inception, the centers have welcomed more than 1.6 million visitors and provided in excess of 800,000 health screenings.

A STRATEGIC PARTNERSHIP

In 1948, Jewish Hospital was designated a teaching hospital, and a formal partnership with the University of Louisville School of Medicine began in 1950. In 1955, the hospital moved into a state-of-the-art facility at Brook and Chestnut streets, adjacent to the medical school, and became an anchor hospital for the Louisville Medical Center.

Jewish Hospital's commitment to partner with the University of Louisville has never wavered, and the Partnership for Progress forged by the two has made a lifesaving difference to the people of this region and beyond. Of the University of Louisville Medical School's 21 departments, Jewish Hospital supports six major clinical affiliations in hand and microsurgery, cardiology and cardiovascular surgery, family and community medicine, organ transplantation, rehab medicine, and plastic and reconstructive surgery.

The Jewish Hospital Foundation, established in 1987, was an early and continuing partner in research and education at the university. Jewish Hospital is a member of the University of Louisville's Minerva Circle, the highest recognition level for contributors of more than $25 million, and the Foundation's Medical Research Grant program has funded a number of noteworthy services.

Since 1996, Jewish Hospital and the University of Louisville have joined in sponsoring RESEARCH!LOUISVILLE, a weeklong conference honoring research initiatives. In 1997, Jewish Hospital partnered with the university and others in a unique concept called Bucks for Brains, a state initiative to jumpstart medical research projects and to attract the best medical talent to the Louisville Medical Center. The program brought Dr. Suzanne Ildstad to Louisville, where she is performing innovative research on transplant patients to help reduce their need for antirejection drugs.

A FOCUS ON THE COMMUNITY

Since its founding in 1903, the bond between Louisville's Jewish community and Jewish Hospital has remained solid. The hospital—one of only four institutions in the country to retain the name Jewish Hospital—and JHHS continue to support the Jewish community in many ways.

Throughout its history, Jewish Hospital has been blessed with exemplary leadership. Talented, involved individuals of all religious faiths continue to serve on the JHHS board, and their dedication has resulted in countless lives being healed and enriched by the resources available at Jewish Hospital and network affiliates.

Significant credit for this success must also be given to President Henry C. Wagner, who for more than 25 years has believed the organization's strength lies in never losing sight of its founding community. His insights and entrepreneurial skills, together with sage board leadership and management expertise, have propelled the organization into becoming a proud network of health care providers with an international reputation for premier medical care.

"Our corporate values," Wagner says, "reflect who we are, have been, and intend to be: a commitment to excellence, Jewish heritage, entrepreneurship, research, education, effective leadership, community service, and respect for all. Our goal has always been to create the best medical address for quality health care."

HENRY C. WAGNER, PRESIDENT OF JHHS, HAS BEEN THE ORGANIZATION'S VISIONARY LEADER FOR MORE THAN TWO DECADES.

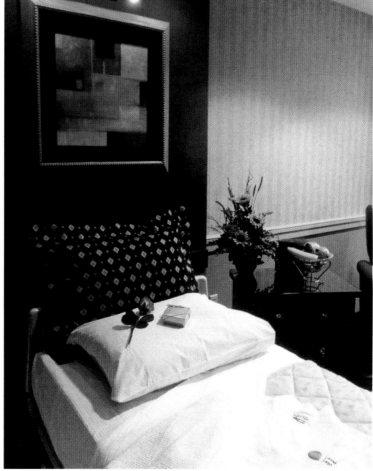

TURNDOWN SERVICE IS ONLY ONE OF THE AMENITIES PROVIDED TO PATIENTS AT THE TRAGER PAVILION, AN EXCLUSIVE AREA THAT COMBINES FIVE-STAR HOTEL ACCOMMODATIONS AND PREMIER MEDICAL CARE AT ONE ADDRESS.

AEGON Insurance Group

AEGON Insurance Group's roots in Louisville date back nearly a century, when Commonwealth Life Insurance Company first became incorporated in 1904. The company sold its first life insurance policy a year later, with a face value of $100 and a weekly premium—collected by an insurance agent—of nine cents. Things have certainly changed since then. The fledgling company originally known as Commonwealth Life is now part of AEGON Insurance Group, the fourth-largest insurance organization in the world, with more than $237 billion in assets and 30,000 employees worldwide. Since its modest Louisville beginnings, the company has become a robust business, a leading employer, a committed corporate citizen, and a vital part of the Louisville landscape.

In 1969, in order to acquire other insurance companies, Commonwealth created Capital Holding Corporation, which quickly evolved from a simple life insurer to a diversified financial services operation of national stature. By 1994, the shareholder-owned company had 10,000 employees nationwide and $24 billion in assets, but its name was not well known. Consequently, Capital Holding renamed itself Providian Corporation, in an effort to create a unified brand for its diverse insurance products and financial services. Then, two years later, Providian became part of the AEGON Insurance Group, headquartered in The Hague, the Netherlands. At the time, it was the largest acquisition in the history of the U.S. insurance industry.

The 35-story AEGON Center is the tallest building in Kentucky, and its lighted, classically domed top easily stands out in the Louisville skyline.

At its landmark headquarters building at Fourth and Market streets in downtown Louisville, AEGON is home to more than 600 employees. With a long-standing reputation as an excellent employer, AEGON was ranked the number-one employer in Louisville in 1998 and 1999 for its family-friendly work policies. These include flexible work arrangements such as job sharing, a company-owned and subsidized child care center, a wellness facility, and many innovative educational and career development programs, in addition to its more traditional benefits. AEGON attracts a diverse, highly expert workforce locally, regionally, and nationally.

Diversified Financial and Insurance Products

More than 85 percent of AEGON Insurance Group's business is in life insurance, pensions, and related savings and investment products. Operating in key markets—the Netherlands, the Americas, the United Kingdom, Hungary, and Spain—AEGON Insurance Group is also present in Belgium, Canada, Germany, Italy, Luxembourg, the Philippines, Taiwan, China, and India. AEGON shares are listed on stock exchanges in six major world financial centers: Amsterdam, Frankfurt, London, New York (NYSE), Tokyo, and Zurich.

Following AEGON's acquisition of Transamerica Corporation last year, several of Transamerica's businesses were integrated into the Louisville operations. AEGON's Louisville offices are engaged in several lines of business, including life and health insurance, fixed and variable annuities, structured settlements, and institutional investment products. Louisville also houses a large investment division, which manages billions of dollars of assets for AEGON's U.S. operations. An industry leader in many product categories, AEGON's products are marketed through a number of distribution channels, including financial institutions, brokers, financial advisors, agents, and marketing organizations, and directly to institutional clients such as 401(k) plans, pension plans, and money market funds. While AEGON is not a household name, AEGON's products touch many individuals and families through retirement and savings plans—either employer-sponsored or individually purchased.

Good Corporate Citizen

Since its inception nearly a century ago, AEGON in Louisville has been a consistent contributor to the city's development. The most obvious contribution, to many Louisvillians, was the addition of the AEGON Center to the Louisville skyline in 1992. The 35-story AEGON Center is the tallest building in Kentucky, and its lighted, classically domed top stands out easily in the Louisville skyline.

AEGON also played a significant role in the renaissance of downtown Louisville in the 1980s, the effects of which continue to spur new downtown growth today. The project known as the Broadway Renaissance was led by Tom Simons, CEO of then-Capital Holding Corporation. His vision was instrumental in the revival of Fourth Avenue and Theater Square, an area that now houses a thriving business and tourism district. An important part of the renaissance plan was the restoration of a magnificent Louisville landmark, the Brown Hotel. Simons' significant contributions to Louisville's development are commemorated in Theater Square, where a bronze statue in his likeness stands in front of the Brown Hotel.

Setting a high standard for corporate giving, AEGON has spearheaded and supported many important community programs. Project Safe Place, a national program that provides shelter for children in trouble, began in Louisville and was primarily funded by AEGON. Today, Project Safe Place is a highly visible program; easily identified by the yellow diamond-shaped signs on participating businesses. Over the years, AEGON has been a major contributor to education and the arts, the Louisville Free Public Library, and a variety of community programs.

Tom Simons' vision was instrumental in the revival of the Fourth Avenue corridor, an area that now houses a thriving business and tourism district.

1911–1955

1913 Ford Motor Company

1916 Peter Built Homes, Inc.

1921 AAF International

1946 Louisville/Jefferson County Metropolitan Sewer District

1948 Manpower Inc.

1950 Anthem Blue Cross and Blue Shield in Kentucky

1950 Bellarmine University

1951 Fire King International Inc.

1951 Prudential Parks & Weisberg Realtors

1953 GE Appliances

1953 Trinity High School

1954 Classroom Teachers Federal Credit Union

1954 L&N Federal Credit Union

1955 Thomas Industries Inc.

Ford Motor Company

Ford Motor Company's history of vehicle production in Louisville began in 1913 on Third Street. After shifting between three other facilities—the Eastern Parkway Plant, Western Parkway Plant, and Louisville Assembly Plant—truck production finally settled at the Louisville Assembly Plant in 1955 and the Kentucky Truck Plant in 1969. The city's geographic location, good workforce, and labor climate made Louisville a natural choice for Ford's new facility.

As the largest truck assembly plant in the world at that time, the Kentucky Truck Plant was a major facility with more than 5 million square feet that occupied 65 acres. Later, it began producing medium and heavy F-Series trucks. The Kentucky Truck Plant produced the Ford Motor Company's line of heavy trucks. The daily production schedule was almost 500 units on two shifts.

The Kentucky Truck Plant produced its 1 millionth truck in October 1979, the beginning of the second decade of the plant's history. At that time, production was more than 500 units per day on two 10-hour shifts.

Unfortunately, economic downturns in the industry beginning in 1980 necessitated the elimination of one production shift. By 1987, the Kentucky Truck Plant had dropped to 2,300 employees producing 300 units per day on one shift. The depressed heavy-truck market continued through most of the 1980s.

EMPLOYEES AT THE FORD MOTOR COMPANY KENTUCKY TRUCK PLANT WATCH THE LAST OF THE L-SERIES TRUCKS ROLL OFF THE LINE IN DECEMBER 1997 (TOP).

THE KENTUCKY TRUCK PLANT IS LOCATED ON 415 ACRES IN EASTERN JEFFERSON COUNTY, KENTUCKY (BOTTOM).

A Decade of Expansion

Numerous changes for the Kentucky Truck Plant occurred in the 1990s. Ford announced that it was transferring its F-Series production from the Kansas City and Ontario truck plants to the Kentucky Truck Plant. In 1993, the plant began producing the F-Series on one shift, and after two years, it was producing the F-Series on two shifts with a line rate of 45 trucks per hour.

The expansion program for F-Series trucks at the Kentucky Truck Plant also produced a modern operation labor agreement, supported by both United Auto Workers (UAW) and Ford management. The F-Series expansion added approximately 1,300 hourly employees and prompted many of the existing heavy-truck employees to move to the F-Series system.

Having produced its last unit in 1997, Ford sold its heavy-truck business to Freightliner. The sale included a substantial portion of the production line's equipment, which was dismantled and reinstalled in various Freightliner facilities. Ford also announced a new program at the Kentucky Truck Plant for the Super Duty F-Series, which required major refurbishment of the abandoned system.

When Ford completely redesigned the F-Series, the Kentucky Truck Plant revised its body shop and underwent major facility changes, adding approximately 900,000 square feet. It insourced crew cab body construction and added the assembly of major closure panels. The program also added the assembly of doors, corner panels, fenders, tailgates, roof bows, and

KELLY BYRD

duce Ford Explorer sport utility vehicles.

PRODUCTION CHANGES

Due to increased customer demand, former heavy-truck employees, new hires, and surplus employees from other Ford locations consistently worked to increase production levels at the Kentucky Truck Plant. The plant's massive training effort for the Super Duty launch involved the merger of the existing F-Series workforce with the heavy-truck workforce, along with the integration of new hires and transfers.

headers that were previously assembled at other suppliers' plants and at Ford's stamping facilities. On January 5, 1998, the all new Super Duty was launched.

Because of the success of this program, Ford selected the Kentucky Truck Plant as the site of a new, integrated stamping facility that would supply parts for the Super Duty truck and plants in Louisville and St. Louis that pro-

With the unbuilt order bank at high levels, Ford converted the Kentucky Truck Plant to a three-crew operating pattern, enabling the plant to grow to 120 production hours each week. As the production hours increased, the three-crew operation pattern required a 50 percent increase in the workforce, consisting mainly of northeast Ohio transfers and new hires.

In 1999, the Kentucky Truck Plant implemented two line-speed increases—from 63 trucks per hour to the present 65 trucks per hour—which were accomplished with only minor capital investments. That same year, the integrated stamping facility started producing inner and outer door panels and Supercab roofs for the Kentucky Truck Plant and Cuautitlan (Mexico) Plant production.

Excursion production began at the Kentucky Truck Plant in July 1999. Although the base Excursion

FROM TOP:
LOUISVILLE DIGNITARIES DRIVE THE JOB 1, COMMERCIAL LIGHT TRUCK OFF THE FINAL LINE IN OCTOBER 1993.

FORD CEO JAQUES NASSAR (LEFT) AND PLANT MANAGER FRANK FOLEY SHAKE HANDS TO COMMEMORATE THE FIRST SUPER DUTY F-SERIES TO ROLL OFF THE LINE IN JANUARY 1998.

EXCURSION PRODUCTION BEGAN AT THE KENTUCKY TRUCK PLANT IN JULY 1999. ALTHOUGH THE BASE EXCURSION PROGRAM WAS DESIGNED TO PRODUCE 10 UNITS PER HOUR, HARD WORK BY EMPLOYEES RESULTED IN A PRODUCTION RATE OF 12 UNITS PER HOUR, REPRESENTING MORE THAN 11,000 ADDITIONAL UNITS ANNUALLY.

HUNDREDS OF KENTUCKY TRUCK PLANT EMPLOYEES DONATED THEIR TIME AND MONEY TO THE WALK TO CURE JUVENILE DIABETES.

PLANT EMPLOYEES PRESENTED MORE THAN $130,000 TO LOCAL CHARITIES AT THE ANNUAL HOLIDAY CELEBRATION.

program was designed to produce 10 units per hour, hard work by employees resulted in a production rate of 12 units per hour, representing more than 11,000 additional units annually.

The Kentucky Truck Plant's mix of regular, super, and crew cabs for a variety of F-Series vehicles—in addition to diesel and 4x4 variants—creates a complex product mix. The plant has one of the most complicated assembly operations in the Ford system. Today, Kentucky Truck Plant covers more than 100 acres under one roof, and employs 6,500 hourly and salaried men and women.

COMMUNITY INVOLVEMENT

Ford is very serious about its responsibilities to the Louisville metropolitan area. The company donates annually to numerous charities, including the Metro United Way, Actors Theater, YMCA, Big Brothers and Big Sisters, and both the Louisville and the Kentucky chambers of commerce. Additionally, Ford has donated used computers to local schools and charities. The company was also a proud sponsor of the 1999 Kentucky Derby. In 2000, Ford Motor Company also co-sponsored Thunder Over Louisville and the Kentucky Derby.

On an individual level, employees at the Kentucky Truck Plant are involved in various community support efforts. They volunteer their time to programs such as Junior Achievement and Green Mile environmental projects. Sixty employees participated in the walk to cure breast cancer in 1998. Habitat for Humanity also benefited from the efforts of Kentucky Truck Plant and Louisville Assembly Plant employees who, in 1998, volunteered their time to build two houses with Ford-donated materials. Ford retirees from both plants have formed a group called Ford ACTS (Assisting the Community through Service). Since its founding in 1990, this group has sorted more than 2 million pounds of food for the Dare to Care Fund.

Ford employees give not only their time to the community, but their money as well. In 1999, they pledged more than $550,000 to the Metro United Way, raised almost $61,000 in a juvenile diabetes walk, and donated more than $21,000 to the crusade for children. At Christmas, employees raised more than $110,000 for the Toys for Tots program, as well as another $32,000 for the Dare to Care Fund.

Combining the latest in production technology and the strength of workforce collaboration, the Kentucky Truck Plant continues to represent a strong Ford presence in the Louisville area. Concern for quality workmanship, as well as community involvement, makes this part of the Ford family a valued corporate citizen.

LOUISVILLE/JEFFERSON COUNTY METROPOLITAN SEWER DISTRICT

Like all river cities, Louisville seems to have a love-hate relationship with the river that runs through it. Without the mighty Ohio, there would be no Louisville, but when the river floods, it has the potential to destroy the city. ■ There is, however, one organization dedicated both to protecting the local waterways and to protecting Louisville from unpredictable flooding—the Louisville/Jefferson County Metropolitan Sewer District (MSD). More than just a sewer service, MSD provides total water quality for all of Louisville and Jefferson County. Its contributions include sanitary sewers, storm-water drainage, and flood protection.

MSD is a nonprofit, regional utility service governed by a board of eight community members. Its main goals are to protect Louisville's streams and groundwater from pollution, and to protect the city's neighborhoods from flooding.

While most customers easily associate MSD with wastewater and drainage surfaces, few realize the vital role the organization plays in flood protection. It is responsible for maintaining and operating the Ohio River Flood Protection System, which includes 15 flood pumping stations and 29 miles of concrete wall and earthen levee. The entire system functioned flawlessly during the flood of 1997. MSD can also provide floodplain information to individuals before they purchase, build, or rent property.

KEEPING WATERWAYS CLEAN

Nonpolluted waterways are a vital part of MSD's mission. To help achieve this goal, MSD connects new customers to sewers each day. This helps to rid Greater Louisville residents of septic tank systems that can lead to the contamination of groundwater. MSD adds approximately 5,000 customers and 90 miles of sewer line each year.

To handle the more than 3,000 miles of existing wastewater sewer lines, MSD continues to update its sewage treatment plants. New technology improves the treatment process and makes the plants better neighbors to the surrounding communities. By 2003, MSD's largest treatment plant, the Morris Forman Wastewater Treatment Plant, will

have in place a new treatment processing system that will better treat wastewater. The new system will also eliminate the odors that result from other wastewater treatment methods.

GREENWAYS AND ENVIRONMENTAL EDUCATION

As part of its commitment to a greener environment, MSD is restoring and reclaiming Louisville's waterways. The community's many creeks and streams are being returned to their original beauty with natural landscaping. This landscaping has a twofold purpose: to bring people back to the natural beauty of the waterway and to provide a natural barrier for pollution.

MSD is also committed to community education as a path to a cleaner environment. The organization has developed a unique partnership with Louisville and Jefferson County schools that focuses on water quality and nature reclamation projects. The partnership includes using high school students to help monitor and analyze water levels, bringing exciting environmental theater presentations to schoolchildren, and inviting students and teachers to its wastewater treatment facilities for tours. In fact, MSD has even set up a special environmental classroom at one of its treatment facilities.

With a forward-reaching goal of improved treatment facilities, responsive flood control, and a cleaner and greener environment, MSD will help Louisville face a bright, clean future in the new millennium.

FROM TOP:
WORKERS BUILD WOODEN FORMS FOR A CONCRETE CHANNEL AT LOUISVILLE/JEFFERSON COUNTY METROPOLITAN SEWER DISTRICT'S NEW, STATE-OF-THE-ART FLOYD'S FORK WASTEWATER TREATMENT PLANT.

AS THE OHIO RIVER FLOODS, MSD WORKERS TIGHTEN THE FINAL BOLTS ON A CLOSURE IN THE DOWNTOWN LOUISVILLE FLOODWALL.

MSD'S GREENWAYS PROGRAM HELPS PRESERVE NATURAL AREAS ALONG THE COMMUNITY'S STREAMS, LAKES, AND PONDS.

Peter Built Homes, Inc.

THE Peter family has long been at the heart of home building in Louisville. Back in 1916, when C. Robert Peter Sr. founded C. Robert Peter Realtors, there weren't as many houses, there weren't as many people, and often there weren't as many regulations designed to protect home buyers. But Peter was a true innovator. He helped make the home building business in Louisville what it is today. He served as the first licensed appraiser in Louisville, assisted in the formation of the Louisville Board of Realtors, and founded Peter Built Homes, seeing it as a perfect complement to his realty business. Eventually he built more than 500 new homes in Louisville.

A Family Affair

This family company has continued to grow since its founding. In fact, C. Robert Peter Jr. is credited with building more homes in Louisville than any other builder. During his lifetime, C. Robert Peter Jr., who joined the family business in 1937, developed more than 28 subdivisions and 2,500 homes. Possessing the same vision as his father, C. Robert Peter Jr. developed homes far from the radius of downtown Louisville, knowing that the outlying areas would be prime land. He helped fuel the quick expansion of Jeffersontown, developing more than 1,500 lots in the area during the 1950s. Some of the Peter Built subdivisions built during the postwar boom include Rosedale, Rosemont, Roselawn, and Charlane Gardens.

The company continued its expansive building toward the east in the 1960s with the Meadowvale subdivision. "At that time, Meadowvale was out in the middle of nowhere," recalls Ben Peter, son of C. Robert Peter Jr. and current owner and president of Peter Built Homes. "Today, it sits in one of Louisville's true hot spots, right on the corner of Hurstbourne Lane and Westport Road."

Ben Peter joined the family company in 1983, and at 19 began developing Harmony Place Condos. During the economic boom of the 1990s, Peter Built Homes continued to develop and build on its tradition of excellence and innovation, responding to the changing needs of the Louisville community. When Ford and UPS expanded their local plants, Peter Built Homes responded by building the Oakhurst subdivision located near the new facilities. "Oakhurst is a 450-lot, master-planned community," says Ben Peter. "Everything about the entire community was planned on paper before it was developed."

Quality, Service, and Innovation

Innovation has always been one of the keys to success for Peter Built Homes. Since 1916, the company has brought numerous firsts to the Louisville home building community. Peter Built Homes was the first company in the area to use drywall, the first to use set-in fireplaces, the first to introduce bi-level and tri-level homes, and the first to build more than 3,000 homes. Peter Built Homes continues

Peter Built Homes, Inc. has long been at the heart of home building in Louisville.

this tradition of innovation today as the first builder to use the squeak-proof Integrity Floor System, as well as being one of the builders chosen for the nationally recognized Heritage Creek subdivision.

Quality and service are critical to the success of Peter Built Homes. It begins with the employees, most of whom have more than 19 years of experience. "We have three employees with more than 30 years of experience, and we have two employees who are the second generation of their family to work for Peter Built Homes," says Ben Peter. "We find good people and we keep them."

Another aspect of quality customer care is the special services that customers can expect after they purchase their home. Home buyers meet with the Peter Built Homes staff in a "pre-framing" meeting and a walk-through before closing. After the sale is complete, the company provides customers with continued support and service. "Usually, when you buy a new home, if you notice a defect you have to go through a complicated and often lengthy process to fix the problem. Instead, we provide all Peter Built Homes customers with 30-day and 11-month checklists that they can mail, E-mail, or fax to us. Within 48 hours of receiving a checklist from the customer, a customer service representative will call and immediately schedule a time for a subcontractor to make the repairs," says Ben Peter.

Since the company's beginning, the Peter family has had a strong sense of community. C. Robert Peter Sr. helped found the Louisville Board of Realtors, and C. Robert Peter Jr. helped found and served as president of the Home Builders Association of Louisville. Today, Ben Peter continues his family's tradition of civic involvement, serving as president for the Louisville Homebuilders Association in 2000.

The Peter family was an integral part of Louisville's growth for nearly all of the 20th century. With a solid family framework and an attention to the needs of its customers, Peter Built Homes will continue to lead the way in providing quality housing to the Louisville community in the 21st century.

PETER BUILT HOMES HAS SET A HIGH STANDARD FOR NEW HOME CONSTRUCTION IN THE LOUISVILLE AREA.

AAF International

Native Louisvillian Bill Reed began his career as a clever entrepreneur operating an automobile paint shop out of his garage. His one nagging problem, dirt settling on his freshly painted automobiles before they had a chance to dry, sent him searching for a solution. Finding no commercially available answer to this dilemma, he developed his own product: a filter consisting of steel wool and chicken wire that he placed on his windows. His invention was so successful that he soon abandoned his automobile painting business and, in 1921, formed American Air Filter, today known as AAF International.

Through continued innovation, Reed led AAF successfully for many years. During the Great Depression, when so many other businesses were failing, his filter company continued to pay dividends. Along the way, Reed and AAF became permanent fixtures in the Louisville business community, and the company became an integral part of the economic foundation of the city. AAF received national recognition when its air filters and manufacturing resources contributed to victory in World War II. The firm's reputation for filtration expertise expanded again when AAF designed and manufactured special filters for use on NASA's Apollo 11 and 13 lunar missions.

The Innovative Spirit Lives On

Today, AAF is one of the world's largest manufacturers and marketers of air filtration products and systems. The company's products now include commercial, industrial, and residential air filters; air pollution control products and systems; and machinery filtration and acoustical systems.

From inexpensive, disposable panel filters to high-efficiency, extended surface filters, AAF markets a wide range of air filters. In fact, the company has developed and introduced most of the filter designs used throughout the industry. AAF air filters can be found everywhere, from manufacturing plants, hospitals, schools, airports, and museums to commercial buildings, hotels, shopping malls, and homes. The firm has been the innovator in the filter business since its inception, and continues to place great emphasis on research and development to meet the increasing demand for cleaner air.

AAF has been marketing its products globally for more than 40 years. From its Louisville headquarters, the company maintains operations in 16 countries with more than 2,100 employees worldwide. Today, AAF brands are recognized throughout the world as the most trusted names in clean air and as symbols of excellence. AAF is supported and enhanced in its international ventures through the resources of its parent company OYL Industries Berhad, located in Malaysia. OYL is a member of the Hong Leong Group, Malaysia, a multinational conglomerate involved in diversified businesses throughout the world.

However, AAF offers the Louisville economy far more than just clean air. In 1999, the AAF corporate headquarters staff was joined by its parent company AAF-McQuay Inc., which employs 5,500 people worldwide and chose Louisville for its new home over several competing cities. The addition of the AAF-McQuay global headquarters brought to Louisville more than 30 top-level positions, which annually bring more than $3 million to the Louisville economy. The two

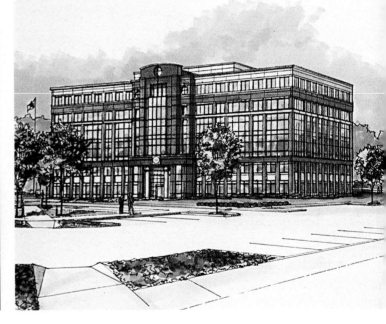

Clockwise from top: American Air Filter, which today is known as AAF International, was founded in 1921.

In 1999, the AAF corporate headquarters staff was joined by its parent company AAF-McQuay Inc., which employs 5,500 people worldwide and chose Louisville for its new home over several competing cities. The addition of the AAF-McQuay global headquarters brought to Louisville more than 30 top-level positions, which annually bring more than $3 million to the Louisville economy.

This AAF OptiFlo cartridge dust collector at a public utility in the United States prevents coal dust from polluting the air.

corporations joined their offices in a new site, the Forest Green office development on Hurstbourne Lane, in May 2000. AAF's sister corporation McQuay International, located in Minneapolis, is an internationally known leader in the manufacture and distribution of air-conditioning and heating products.

21ST-CENTURY TECHNOLOGY

The need for clean air is universal and unlimited, and transcends political environments, economic uncertainties, and markets. The Environmental Protection Agency ranks indoor air as one of the top five environmental threats to human health. According to the World Health Organization, 30 percent of commercial buildings show signs of sick building syndrome, an illness caused by exposure to pollutants or germs inside an airtight building. As evidence of how pervasive this problem is, indoor air quality has been established as the number one management issue by the International Facility Management Association. While many factors contribute to poor indoor air quality, the largest contributor is particulate dust and microbial contamination.

AAF is taking on this 21st-century problem with true 21st-century technology, and is a recognized leader in developing filters that improve the quality of indoor air. AAF has improved the ability of its filters to capture particulate dust by developing new filtration media and employing innovations in filter design. In industrial settings, this new technology can be lifesaving, and in residential uses, these filters improve the air for home owners and their families.

In addition to developing its own revolutionary products, AAF has been participating with Ohio State University and the Cleveland Clinic in a study on methodologies for effectively reducing allergens in the home. The company is applying the results of this research to the development of new products.

AAF has also developed modern filters to meet the demands of today's increasingly high-tech society. For example, the company has pioneered many of the techniques and filter products used in clean room operations. In addition, AAF engineers have played an active role in establishing new standards for clean room applications.

BETTER AIR IS OUR BUSINESS

A Louisville-grown company, AAF continues to remain an important part of the Louisville economy. Worldwide, the company's focus is the air filter business. Keeping its founder's innovative spirit, AAF will continue to provide the world with clean air solutions for decades to come. The company's motto is the same today as it was when Reed constructed his first filter: "Better Air is Our Business." And at AAF, business is improving each day.

AAF MANUFACTURES AN EXTENSIVE LINE OF AIR FILTERS.

AAF DESIGNED A SPECIAL FILTER TO PREVENT CONTAMINATION OF THE MOON BY EARTH PARTICLES DURING THE APOLLO LANDINGS.

MANPOWER INC.

With more than 2 million employees worldwide, Manpower Inc. is the world's leading provider of staffing and workforce management solutions. The company has a network of 3,400 offices in 53 countries and services 400,000 businesses worldwide. In addition, 96 percent of the Fortune 500 companies rely on Manpower to assist with their staffing needs. Founded in Milwaukee in 1948, Manpower provides customized, innovative staffing solutions and services to businesses across the globe. "In Louisville, we work closely with hundreds of leading businesses to ensure that they have the skilled staff to support their operations," says Sherry Neisen, area manager for the Louisville Manpower office.

"Manpower became a part of the Louisville business community offices in the surrounding area," says Neisen. "We support businesses of all types, including general office, call center, light industrial, financial, and information technology organizations. Manpower provides a full spectrum of services such as supplemental staffing, on-site management programs, and direct hiring for permanent placement. Furthermore, Manpower can play a role in an outsourcing capacity by acting as an extension of a company's human resources department."

With historically low unemployment rates and increasing global competition, more and more companies are asking Manpower to help them identify and implement effective workforce solutions directed at their strategic objectives.

MAKING THE RIGHT MATCH

To ensure that each of its employees will always be placed at the right job, Manpower utilizes its proprietary Predictable Performance System, which assesses each applicant's knowledge, skills, abilities, and aptitude. In addition, the company performs on-site analyses of its clients' facilities to gain a full understanding of the job responsibilities required, work procedures followed, and work environment involved. This proven method provides the best possible match between Manpower's employees and its customers.

Manpower can also customize its testing and training programs to specifically meet the business needs of its customer. For example, Manpower can create customized

Through Manpower Inc.'s on-line university, the Global Learning Center, the company's employees have 24-hour access to more than 1,000 skills training courses, assessment tools, and management services.

Manpower, the world's leading provider of staffing and workforce management solutions, has 3,400 offices in 53 countries.

software training applications to match a company's existing system, and can then use this system to train Manpower employees so they are fully productive upon placement at the company.

PROFESSIONAL STAFFING

To meet the growing demand for high-end professional staffing solutions, the firm has a dedicated operating unit called Manpower Professional. This unit specializes in the assignment of information technology, financial, telecommunications, scientific, engineering, human resources, and quality professionals. There are more than 200 Manpower Professional offices worldwide.

GLOBAL LEARNING FOR THE WORKFORCE

A key element to Manpower's success is the extensive array of free training the company offers its employees. Training is one of the major reasons why people work for Manpower and why businesses seek the firm's services.

With its on-line university, the Global Leaning Center (GLC), Manpower keeps its employees on the cutting edge of the latest technologies and industry information. This Internet-based training and career management program offers skills assessment, professional certification information, and other career services, including more than 1,000 courses in the most popular software programs, information technology, and business skills.

Manpower also offers its employees computer-based training in the latest software applications through its Skillware training program. Skillware has equipped workers worldwide with the skills to use word processing, spreadsheet, database, presentation graphics, and other programs. More than 75 of the Fortune 100 companies have utilized Skillware to train their employees.

To meet the increasing demand for information technology professionals, Manpower has also developed TechTrack, which includes more than 1,000 courses in today's most in-demand networking, programming, and client-server technologies such as Java, SAP, and Windows NT. Manpower also administers certification testing in a wide variety of programs including Microsoft, Novell, and Unix.

Manpower provides professional development and skills training with its interactive PowerSkills series. The program teaches essential skills such as negotiation, creative thinking, improved decision making, and business strategies.

THE TOTAL SOLUTION

As an important part of its successful philosophy, Manpower reaches out to people in the community who at times are underutilized in the workforce, such as military veterans, people with disabilities, older workers, and people transitioning from military to civilian life. "We're in the business of putting people to work," says Neisen. "Manpower's assessment tools not only reveal what a person's strengths are but also uncover a person's hidden talents."

Manpower continually ranks as the leading staffing firm in the world. In addition, each quarter it conducts an Employment Outlook Survey, a nationwide survey of the permanent hiring plans of more than 16,000 companies in 485 cities. Government leaders, news media, and economists use the survey as a leading indicator of labor market trends.

"Manpower is committed to providing exceptional service to the business communities in Louisville and the surrounding area," says Neisen. "By utilizing our proven recruitment, assessment, and training tools, we provide the best staffing solutions for both our employees and our customers."

THROUGH A COMPREHENSIVE SYSTEM OF ASSESSMENTS, SUCH AS ITS ULTRADEX TEST FOR LIGHT-INDUSTRIAL WORKERS, MANPOWER IS ABLE TO SUCCESSFULLY MATCH THE RIGHT WORKER TO THE RIGHT JOB IN A WIDE VARIETY OF FIELDS.

MANPOWER PROVIDES A FULL SPECTRUM OF SOLUTIONS AND SERVICES TO SUPPORT THE STAFFING NEEDS OF MANY BUSINESSES, INCLUDING CUSTOMER SERVICE, OFFICE AND CLERICAL, LIGHT INDUSTRIAL, HUMAN RESOURCES, FINANCIAL, AND INFORMATION TECHNOLOGY.

Anthem Blue Cross and Blue Shield in Kentucky

The history of Anthem Blue Cross and Blue Shield in Kentucky begins more than 60 years ago with the formation of the Louisville Community Hospital Service. This early health plan was created to help make hospital stays more affordable and quickly won the endorsement of the American Hospital Association. In 1939, the company began employing the Blue Cross name and emblem.

From those modest beginnings eventually grew the Southeastern Mutual Insurance Company, the Blue Cross and Blue Shield licensee for Kentucky. Consolidation with Blue Cross and Blue Shield-licensed companies in Indiana and Ohio in 1993 and 1995 created what is now known as Anthem Blue Cross and Blue Shield. Today, the Anthem Blue Cross and Blue Shield companies in the Midwest include Anthem Health Plans of Kentucky, Inc., Anthem Insurance Companies, Inc. in Indiana, and Community Insurance Company in Ohio. Each company operates as Anthem Blue Cross and Blue Shield in its respective state. Headquartered in Indianapolis, the Anthem family of companies is a strong yet flexible, growing yet community-focused leader in health care management. The Anthem mission, pursued through its strong provider networks, innovative benefit programs, and health education initiatives, is "to improve the health of the people we serve."

Local Service

Locally Anthem provides insurance coverage for more than half a million people in Kentucky. However, in order to offer access to effective care, the company has created 12 individual health service areas (HSAs) in Indiana, Kentucky, and Ohio. This system enables Anthem to assemble within each region a network of hospitals, doctors, and other caregivers designed to serve Anthem members.

Sixteen hundred Anthem associates work at the company's Louisville location. They strive to improve service within Kentucky's local HSAs by periodically updating and refining Anthem's network of physicians and by offering programs and benefit plans that help educate members and promote healthy lifestyles.

Blue Preferred® Primary, a health maintenance organization (HMO); Blue Preferred® Primary Plus, a point-of-service (POS) program; and Blue Access℠, a preferred provider organization (PPO), are Anthem's major managed care products in Kentucky. This variety of programs allows customers to choose the right products to match their individual needs. All Anthem products, in keeping with the goal to help its members live healthy lives, include coverage for numerous preventive services such as well-baby care, immunizations, and routine health screenings.

National Strength

While enjoying the benefits of convenient and responsive local health care networks, Anthem members also get the peace of mind that comes with being associated with a large, stable institution. In recent years, the Anthem family of companies has expanded to include Anthem East, with Blue Cross and Blue Shield-licensed companies in Connecticut, Maine, and New Hampshire, and Anthem West, which serves Colorado and Nevada. Anthem has carefully studied the strengths of each of its new additions and is striving to successfully incorporate those best qualities into the overall organization. For example, Anthem Blue Cross and Blue Shield of Connecticut had a highly informative and easy-to-navigate Web site. The site has now become the model for all of Anthem's companies.

During its expansion, the Anthem family of companies has maintained its sound financial standing. The company's surplus exceeds $1.7 billion, and its current industry rankings from independen

Anthem Blue Cross and Blue Shield has been serving the Louisville community for more than 60 years.

ratings agencies include an A- (Excellent) rating from A.M. Best Company; an A rating from Standard & Poor's for claims-paying ability; and an A+ rating from Duff & Phelps in the same category. Anthem intends to continue to build on this solid, secure base.

Kentucky's Anthem members also get the benefit of the BlueCard and BluesCONNECT® programs. With BlueCard and BluesCONNECT, customers can get access to health care while away from home by simply showing their Anthem Blue Cross and Blue Shield identification cards to participating providers.

Building Healthy Communities

Anthem and its associates use three guiding principles for its community service programs: Healthy Minds, in support of education; Healthy Bodies, promoting health and fitness; and Healthy Communities, encouraging economic opportunity and the arts, as well as funding social services.

Anthem's Healthy Woman initiative showcases the company's educational efforts. This ambitious program focuses on health issues of special concern to women, including high-risk pregnancies, hormone replacement therapy, and the prevention and detection of heart disease, cancer, osteoporosis, and other conditions.

To encourage fitness, Anthem Health Plans of Kentucky, Inc., along with the *Courier Journal*, sponsors the annual Kentucky Derby Festival Mini-Marathon. And, in summer 2000, Anthem, along with the other independent Blue Cross and Blue Shield Association member plans, brought the Blue Cross and Blue Shield Caring for the Human Spirit™ tour to Louisville. This event, part of Blue Cross and Blue Shield Association's sponsorship of the U.S. Olympic team, included demonstrations of sports, such as judo, and exhibits on nutrition and training. The event's mini-decathlon gave armchair athletes a chance to test their skills and fitness against potential medal winners on the U.S. team.

Each year, Anthem and the American Cancer Society conduct a free skin and oral cancer screening at the Kentucky State Fair. In 1999, more than 20,000 people were checked. This service, along with generous contributions to the United Way campaign and the Fund for the Arts, demonstrates Anthem's commitment to helping Louisville remain a strong, vibrant, and healthy community.

ANTHEM PROGRAMS ENCOURAGE EDUCATION AND DISEASE PREVENTION (LEFT).

EACH YEAR, ANTHEM COSPONSORS THE KENTUCKY DERBY FESTIVAL MINI-MARATHON (RIGHT).

APPROXIMATELY 1,600 ANTHEM ASSOCIATES LIVE AND WORK IN LOUISVILLE.

Bellarmine University

In October 1950, Bellarmine University first opened its doors under the sponsorship of the Roman Catholic Archdiocese of Louisville with the special assistance of the Conventual Franciscan Fathers. Originally a college for men only, Bellarmine merged in 1968 with Ursuline College, a Catholic school for women established by the Ursuline Sisters of Louisville in 1938.

Becoming coeducational in 1968, Bellarmine is today an independent college with a self-governing board.

The new millennium carries Bellarmine University into the second half of its first century serving Louisville and the surrounding region with an educational environment of academic excellence in the Catholic liberal arts tradition. At its inception, Bellarmine was one of the first colleges in the Commonwealth of Kentucky open to all races. The University continues to attract talented and diverse people of all faiths and ages to develop the intellectual, moral, and professional competencies needed to lead, to serve, to make a living, and to make a life worth living.

Educational Leader

Its distinguished educational tradition and commitment to academic excellence have helped Bellarmine University achieve a ranking as one of the top 25 regional universities in the South by *U.S. News & World Report*. Bellarmine offers students a well-rounded academic program with more than 40 different majors. And the more than 2,500 students at Bellarmine can continue their education in one of three different graduate degree areas: business, nursing, and education.

"Although our primary goal has always been to help students develop their God-given gifts, we have always recognized a broader mission to the community," says Dr. Joseph J. McGowan Jr., president of Bellarmine University. "Community service and service learning are a central part of the Bellarmine experience, inside and outside the classroom. In turn, Bellarmine students benefit from the resources of the city of Louisville through a wide variety of internships with local employers preparing students for their careers."

Combining academic life with student activities is also an important part of the Bellarmine experience. There are more than 50 different organizations, sports teams, and clubs for students to choose from. One shining example of the success of the college on all levels is the Bellarmine mock trial team, which won the national championship in 1999, competing against teams from Harvard, Yale, Columbia, Cornell, Princeton, and many others.

Bellarmine has grown to meet the changing needs of its increasing student population. Today, the college has nearly 20 campus buildings on more than 120 acres of gently rolling terrain that slopes down to Louisville's historic Beargrass Creek.

Throughout the first part of the new millennium, Bellarmine is anticipating continued growth, and already has the funds and plans for several new building projects. With more residential housing units and additional classroom space, Bellarmine will nearly double its current student population, moving it to the forefront as one of the best private Universities in the region.

Its distinguished educational tradition and commitment to academic excellence have helped Bellarmine University achieve a ranking as one of the top 25 regional universities in the South by U.S. News & World Report. Bellarmine offers students a well-rounded academic program with more than 40 different majors.

GE APPLIANCES

The GE Appliances global headquarters in Louisville is destined to become the place of origin for the next generation of smart appliances. Home to a dedicated and talented group of engineers and other high-technology professionals, the facility's main purpose is to create and design innovative and extremely reliable products with the help of world-famous Six Sigma quality focus. ■ Today's consumers want reliable, smart appliances that make their lives easier. GE Appliances' innovative corporate team created more than 20 new products in 1999 and more than 40 in 2000 to meet this demand. The global headquarters directs many of the worldwide functions for GE Appliances, including the development, testing, worldwide sales and service, and marketing of these new products.

GE Appliances is dedicated to the city of Louisville and continues to be a global high-technology leader. GE has led the way in the appliance industry to invest in E-commerce. The company is a Web-centric place with scores of Six Sigma experts who tear down up to 100 Web sites every night while the city sleeps, and reinvent GE's consumer Web site every morning. GE may be Kentucky's largest dot-com. With its Sigma focus on innovative and dependable products, GE will help Louisville usher in its new position as a major technology center in the new millennium.

SIX SIGMA SUCCESS

Six Sigma is a highly effective methodology, led and taught by highly trained GE employees called master black belts. Although the average consumer has probably never heard of Six Sigma, this methodology is improving every GE product and service on the market.

The methodology focuses on improving products and delivering what the customer wants in a measurable way. For example, the GE Triton dishwasher was the first to be fully designed under Six Sigma technology. It was engineered and designed in half the time of previous dishwashers. The Triton meets the expectations of the most discriminating consumer and has become known as "the quietest dishwasher in America."

WORLDWIDE SERVICE CENTER

Another important part of the GE Appliances global headquarters is the worldwide service centers that are housed in the facility. Probably the most well known of these is the GE Answer Center, which answers nearly 2,000 consumer E-mails per week and 2 million calls per year. Known as the gold standard in the industry for call-taking centers, GE's Answer Center continues to impress customers and industry analysts alike with its speed and efficiency. In fact, the average waiting time for an E-mail response from the GE Answer Center is 10 minutes, compared to many others in the industry who take days to respond to E-mailed questions.

The Internet has also become a new focal point for worldwide sales of GE appliances. Currently, there are sites where individual builders and appliance dealers can order all GE appliances on-line. Individual consumers can order on-line through a dealer Web site or by finding a dealer through www.geappliances.com; a local dealer will then provide the delivery, installation, and service. GE is the first appliance manufacturer in the industry to provide wireless phone or personal digital assistant (PDA) access to its customer Web site for dealers and builders. The ease of ordering products on-line makes GE an E-commerce leader in the appliance industry, and is one of the reasons *Internet Week* recognized GE with its E-Business of the Year award.

FROM ITS HEADQUARTERS IN LOUISVILLE, GE APPLIANCES HAS BECOME A HIGH-TECHNOLOGY HOTBED FOR THE DEVELOPMENT OF INNOVATIVE PRODUCTS.

Fire King International Inc.

Founded in 1951 by L.G. Carlisle, Fire King International Inc. is the nation's leading manufacturer of insulated, fireproof storage systems. ■ Over the past 25 years, Fire King International Inc. has maintained significant growth and established a solid foundation as a strong competitor in the fireproof storage industry. In 1974, Van Carlisle, CEO of Fire King International Inc., took over his family's business following the death of his grandfather, L.G. Carlisle, in 1973. Fire King International Inc. was losing money and ranked last in a field of six producers of fireproof filing cabinets. At 25 years of age, Carlisle was determined to establish Fire King International Inc. as an industry leader.

Carlisle managed to focus his company on providing superior products, offering only high-quality fireproof cabinets, while others in the industry offered a lower-priced but less effective fireproof cabinet. Through hard work and dedication, Carlisle was able to develop a complete line of high-quality security products, ranging from fireproof file cabinets to high-security safes to digital surveillance systems.

Today, Fire King International Inc., located in New Albany, Indiana, employs more than 600 people and is the number one producer in the fireproof file industry with more than 70 percent market share.

A Company on Fire

In 1974, Fire King International Inc. was already the industry leader in quality. Carlisle understood the need to convince his dealers of the importance of quality and its relationship to Fire King's product line.

Because Fire King's products were sold primarily through wholesalers and dealers, Carlisle developed several strategies, including a prepaid freight program, to heighten the dealers' interest in his company's product. His approach worked, and the company began to grow at an extraordinary pace, earning Carlisle the title of Master Strategist by *Inc.* magazine.

Fire King's products continue to outdistance the competition in quality. The company's file cabinets carry the stringent Underwriters Laboratories Class 350 one-hour fire and impact rating, meaning they have passed a 1,700-degree fire endurance test and a 30-foot drop/impact test. Fire King is the only manufacturer to carry a file cabinet rated with a two-hour fire rating.

Safe and Secure

Understanding the need to incorporate fireproof safety into additional products, Fire King expanded its line in 1990 by acquiring Meilink, a manufacturer of safes. "We saw a real need to provide business clients with better safes," says Carlisle. Operating as a separate division of Fire King, Meilink has been an industry leader in the safe market for more than 100 years. Today, through Fire King, Meilink offers a safe for every security need.

Fire King has long been known as a client-focused manufacturer. During the 1990s, Fire King's client-focused strategy led to the creation of a new product. At the request of one of its largest safe customers, Tricon Global Restaurants, Fire King—in partnership with NKL Safe Company—developed the first "smart safe." The result was a safe with an electronic lock that provided a complete audit trail of every safe use.

Today, NKL Safes, now a division of Fire King, provides computerized functions helping businesses ensure greater security. The safes

Van Carlisle is president and chief executive officer of Fire King International Inc.

can automatically count the deposited money, provide a printer interface and encrypted audit trails, and limit access through digital access, user PIN numbers, custom time locks, and custom time delays. They will even notify authorities when they are opened during nonbusiness hours or when an employee must perform an emergency opening during a robbery. Providing far more than just security for money, these safes improve the bottom line and are helping to protect lives as well.

Fire King also offers a wide variety of safes for use in residential applications. An important part of the residential line of safes is Winchester Gun Safes. These safes come in a variety of designs made to hold handguns, shotguns, and household valuables, keeping them safe from fire, children, and burglars.

THE FUTURE IS SAFER

Fire King continues to develop proprietary electronics, enabling the company to remain on the cutting edge of technology. By increasing standards and production capabilities, Fire King continues to provide the highest-quality fireproof filing cabinets and safes possible. In 2000, Fire King invested more than $8 million in technology and equipment upgrades and expansions to its manufacturing headquarters in New Albany. This new technology allows Fire King to offer additional high-technology equipment to the market.

THE "SMART SAFE," AN EXAMPLE OF FIRE KING'S INNOVATIVE STYLE THAT HAS CONTINUED TO GROW INTO NEW MARKETS, FEATURES AN ELECTRONIC LOCK TO ENSURE BUSINESSES GREATER SECURITY BY PROVIDING AN AUDIT TRAIL OF EVERY SAFE USE.

A new product designed to increase safety is Image Vault, a multifunctional, digital recording system. The idea of Image Vault was transformed from a customer's request for Fire King to help it improve the videotape closed circuit television security system. Image Vault replaces the traditional time lapse VCR equipment by providing a higher-quality image. Because of Image Vault's accuracy and remote access features, it improves security and offers many operational benefits.

"Partnering with our customers to solve their security problems is what makes us the innovators of the marketplace," says Carlisle. Fire King is also developing new products, including such 21st-century features as biometrics, which use the human body's imprints—retinal scans and fingerprints—as part of its user identification protocol.

Twenty-five years ago, only one person believed Fire King would be the industry leader in sales and quality, and would be an innovator in incorporating the latest technologies into its products. That belief, backed by hard work, proved to be right. "I knew all along Fire King had the best product," says Carlisle. "My challenge was teaching both the distributors and the customers to notice our quality products."

FIRE KING RECENTLY EXPANDED ITS HEADQUARTERS AND DISTRIBUTION CENTER IN NEW ALBANY, INDIANA.

PRUDENTIAL PARKS & WEISBERG REALTORS®

PRUDENTIAL PARKS & WEISBERG REALTORS® HAS BEEN A LEADER IN the Louisville-area real estate market since 1951. The company was originally known as Bass & Weisberg, but in 1980, after nearly 30 years in the business, Louis Bass retired and sold his portion of the company to the Weisberg family. Despite these changes, the company's dedication to quality customer service has never faltered.

Today, cofounder Charles Weisberg's sons remain very involved in the business and are the owners: Frank Weisberg serves as chairman and Ron Weisberg is executive vice president. Frank, the first Certified Commercial Investment Member (CCIM) in Kentucky and a national educator, and Ron, a Certified Property Manager (CPM) honored as CPM of the year, are both committed to quality and education for the company's agents and staff. Since the mid-1990s, the firm has been affiliated with the largest financial group in the nation, Prudential. Prudential Parks & Weisberg Realtors® remains a locally owned company that is able to offer the benefits of a nationwide real estate firm.

PRUDENTIAL PARKS & WEISBERG REALTORS® TEAM INCLUDES (FROM LEFT) FRANK WEISBERG, CHAIRMAN; JUDIE PARKS, VICE PRESIDENT OF BUSINESS DEVELOPMENT, CHAIRMAN'S CIRCLE; PAT PARKS, PRESIDENT; ELLEN SHAIKUN, CHAIRMAN'S CIRCLE; AND RON WEISBERG, EXECUTIVE VICE PRESIDENT.

HAROLD AND BONNIE COHEN HAVE BEEN TWO OF PRUDENTIAL PARKS & WEISBERG REALTORS' TOP PRODUCERS SINCE 1985.

QUALITY LEADERSHIP AND TRAINING

In 1984, the Weisbergs promoted Pat Parks to the office of president. Known for her vision and business acumen, Parks continues to serve as president and COO. Often nationally recognized as an outstanding Realtor and speaker, she was the first person in the region to serve as the national president of the Women's Council of Realtors. Parks' management and training expertise has helped the company become one of the largest privately owned, full-service firms in the region.

Parks began her career with the company in 1974 as a training director and Realtor. As her career path has developed, Parks has consistently focused on three main areas: training Realtors, improving service, and giving back to the community. She helped create the national Realtor-training program *Leadership Training Graduate*, and continues to serve as a trainer, offering seminars and development programs to Realtors nationwide.

Parks provides a comprehensive training program for her own staff. For many Realtors, it is the quality of these programs that brought them to Prudential Parks & Weisberg Realtors. "Out of necessity, I changed careers mid-life and I didn't have years to spend training," says Bonnie Cohen, a 15-year top producer for the firm. "I needed to make money immediately. Pat knows what she is doing and she can help agents succeed right away."

The ever changing regulations and new technologies that affect the field of real estate make up-to-date training a necessity. The firm's long-term Realtors, such as Cohen, continue to receive training throughout their careers.

FULL-SERVICE REAL ESTATE

As a true full-service real estate company, Prudential Parks & Weisberg Realtors can fulfill every real estate need for its clients. The firm offers residential and complete relocation services, property management and business services, commercial services, and its own mortgage company, Multiple Options Mortgage Services.

Prudential Parks & Weisberg Realtors attributes much of its success to the longevity of the company and the vast experience of its Realtors. "Because we have been here for so long offering quality service, we have many repeat customers, through referrals or even through multiple generations of families," Parks says. "In situations where we have worked

with grandparents or parents, the children are now buying homes with our help."

Relocation services are another specialty of Prudential Parks & Weisberg Realtors. For people who are being relocated to Louisville, the agents not only help find the right home, but also help locate schools, provide a job bank to help a spouse find a job, and find temporary rentals if needed.

Advancing with Technology

As technology changes, Prudential Parks & Weisberg Realtors is dedicated to offering the latest advancements for its customers. "We are listing all of our homes, rentals, and commercial properties on the Internet," Parks says. "Buyers can actually log onto our Web site, www.pwprudential.com, and create a profile that fits their purchasing needs. Based on their own criteria, they can browse our current listing of homes fitting those parameters. In the past, it was difficult for buyers to receive current information." The firm's Web site information is updated every day.

In addition to using the Internet, Prudential Parks & Weisberg Realtors also provides a detailed property source service with current home listings, available by phone or fax. As an added advantage, the firm's Realtors use the latest services offered through Prudential, including value range marketing, which allows homes to be listed under a price range instead of one set price. More than half of Prudential Parks & Weisberg Realtors' agents are certified, showing expertise in today's technology.

Community Involvement

Prudential Parks & Weisberg Realtors prides itself on a high level of community involvement, and the top management team is made up of Louisville natives who are committed to community service.

The firm's staff and Realtors are encouraged to participate in a variety of service programs. Each year, Prudential Parks & Weisberg Realtors presents an award to the staff member who is the most active in the community. The company supports Sunshine Kids, a Prudential program affiliated with Kosair Children's Hospital that provides recreational activities and vacations for children who have cancer.

"Every aspect of real estate is about helping people, and this field attracts people who want to be involved in the community and want to help people live better lives," Parks says. "That is what real estate is all about."

Chairman Frank Weisberg speaks to the company's top producers with Louisville in the background.

The company's Prudential national award winners include (from left) Bill Tindell, Karen Foster, Judie Parks, Ruth Anne Thompson, Bonnie Cohen, Beth Cress Rose, Ellen Shaikun, Al Fowler, Pat Parks, Frank Weisberg, and Ron Weisberg.

Trinity High School

At Trinity High School, a tradition of academic excellence and technology collide. In its almost 50-year history, this Catholic high school has been a consistent innovator in the education of young men. ■ Trinity's commitment is to welcome all students regardless of academic, religious, or racial background, and to afford each student every opportunity to reach his potential.

With this belief, Trinity has programs for all levels of students, including advanced programs for gifted students, as well as programs designed for students with learning differences. The majority of students attending Trinity are enrolled in a college preparation curriculum, and more than 95 percent of Trinity graduates go on to college annually.

Strong Catholic Tradition

Trinity High School first opened its doors in 1953. Louisville Archbishop John Floersh anticipated the growth of Louisville's eastern suburbs, and chose the site of a former Catholic church and school in St. Matthews as the location for a new Catholic high school. With the Catholic tradition of teaching and learning firmly embedded in the buildings that became Trinity, the school is now more like a small college campus with more than 1,100 students. Trinity is conveniently located in a highly sought after neighborhood, easily accessible to all parts of the metropolitan area.

Today, the school boasts 10,500 alumni, and has been nationally recognized as a Blue Ribbon School of Excellence by the U.S. Department of Education. Students have gone on to distinguished careers as scientists, doctors, lawyers, teachers, ministers, professional athletes, skilled tradesmen, and entrepreneurs in every imaginable field and endeavor in the public and private sector.

Growing Character the Trinity Way

Trinity strives to educate young men for life, believing the school's purpose is to foster the personal growth and maturity necessary to live a successful Christian life in a complex society. As a Catholic school, Trinity provides the moral fabric for character formation based on Christian values in the Catholic tradition. As an example, through Trinity's Community Service Program, students are expected to contribute a specified number of service hours to the community, helping families, elementary schools, nursing homes, and many community organizations. Schoolwide worship experience, retreats, and days of recollection add to the gospel-centered, mission-driven focus.

The relationship formed with teachers is an integral part of character formation. With a low

THE RELATIONSHIP FORMED WITH TEACHERS IS AN INTEGRAL PART OF CHARACTER FORMATION. WITH A LOW STUDENT-TO-TEACHER RATIO, TRINITY'S DEDICATED AND CARING FACULTY ENCOURAGES STUDENT PARTICIPATION AND INPUT (LEFT).

TRINITY STRIVES TO EDUCATE YOUNG MEN FOR LIFE, BELIEVING ITS PURPOSE IS TO FOSTER THE PERSONAL GROWTH AND MATURITY NECESSARY TO LIVE A SUCCESSFUL CHRISTIAN LIFE IN A COMPLEX SOCIETY. AS A CATHOLIC SCHOOL, TRINITY PROVIDES THE MORAL FABRIC FOR CHARACTER FORMATION BASED ON CHRISTIAN VALUES IN THE CATHOLIC TRADITION (RIGHT).

student-to-teacher ratio, Trinity's dedicated and caring faculty encourages student participation and input. The school's philosophy states, "The student is the central focus of Trinity." This approach, combined with an emphasis on spiritual and personal growth, results in students who consistently achieve their potential.

Getting involved is easy for students at Trinity. There is a wealth of co- and extracurricular activities to educate for life. Annually, students from Trinity receive awards for excellence in drama, journalism, speech, art, music, and academic competition.

The Trinity tradition is carried on in athletics as well. Trinity holds the most state football championships in the toughest class in Kentucky. The school has also won multiple state titles in cross-country, golf, swimming, volleyball, tennis, soccer, track, and wrestling. No school of like age in the Louisville area has equaled Trinity's level of success.

Cyberschool

Trinity has positioned itself as Louisville's cyberschool. In the past four years, Trinity has allocated nearly $1 million from institutional resources to advance the school's technology capabilities. All students have easy access to computers and can even check out a laptop overnight.

Trinity's integrated school network has Internet access in all classrooms and a schoolwide E-mail system. Trinity is one of the first high schools in Louisville to have its own Web page, located at www.thsrock.net and the only one to have a webmaster on staff. With technology as the focus of curriculum, there are now five full-time instructors and coordinators dedicated to expanding Trinity's cyberschool reputation.

Trinity has already put its advancements to good use in the community. In the World Wide Web design class, students constructed and now maintain the Kentucky Derby Museum's award-winning Web page.

Innovation and Self-Determination

In 1993, Trinity became an independent, self-governing school operating under a sponsorship agreement with the Archdiocese of Louisville, creating a unique model of president-principal administration for the nation. The future of Trinity is sustained not only by its dedicated mission, but also by an enormous amount of support from the community and alumni. Thanks to this support, Trinity now has a growing endowment that is managed by its own Trinity Foundation Board of Directors.

Trinity is the first Catholic high school in Louisville to have its own site-based school board. There are three volunteer organizations working toward maintaining the Trinity tradition: the Trinity High School Board, the Alumni Board, and the Foundation Board. These boards oversee the 15-Year Vision, including a series of three-year strategic plans guiding the direction of the school. It is clear that Trinity is shaping its own future, not simply reacting to what comes next.

The Future Looks Bright

The Trinity tradition, with its unique spirit and values, continues to prepare young men for success in the next century. Although technology will change many things, a remaining constant will be the Catholic-faith-based core values that Trinity instills in its students.

The students at Trinity—present and past—are proud of their connection to the school and to each other. A Trinity alumnus puts it all together: "Every teacher influenced me in some way. Academically, the door was always open to succeed. Trinity helped me become the person I wanted to be academically, physically, and spiritually. I would not be where I am now without Trinity's gift, which gave me the discipline to stay focused on my dreams."

TRINITY HOLDS THE MOST STATE FOOTBALL CHAMPIONSHIPS IN THE TOUGHEST CLASS IN KENTUCKY (TOP).

TRINITY FEATURES A VARIETY OF SPECIALIZED PROGRAMS TO PREPARE STUDENTS FOR SUCCESS IN THE WORLD (BOTTOM).

Classroom Teachers Federal Credit Union

CLASSROOM TEACHERS FEDERAL CREDIT UNION (CTFCU) was formed in 1954 for the employees of the Jefferson County Board of Education. Initially, the credit union offered only limited services; the board of directors adopted bylaws that allowed members to borrow no more than $100 on an unsecured loan. Today, CTFCU members can take out loans for a wide variety of amounts to cover everything from vacations to new homes. Nearly 50 years after its inception, CTFCU members have secured a total of $192 million in loans, while the credit union's assets have grown to more than $60 million.

Although CTFCU originally enrolled only employees of the Jefferson County Public School System, the credit union now services 28 other educational institutions in the Louisville metro area. The membership includes employees of the Bullitt, Oldham, Shelby, Spencer, and Nelson county public school systems, as well as many private elementary and high schools; business colleges and technical schools; Kentucky Educational Television (KET); and University of Louisville students and alumni. Enrollment is open to all employees at these educational institutions—from the lunchroom worker to the superintendent—and their families.

CLASSROOM TEACHERS FEDERAL CREDIT UNION (CTFCU) RECENTLY FINISHED THE RENOVATION OF ITS MAIN OFFICE, REMODELING 8,450 SQUARE FEET OF OFFICE SPACE AND COMPLETING A 12,000-SQUARE-FOOT, TWO-STORY ADDITION. ONE OF THE PRIME FEATURES OF THE NEW BUILDING IS AN INTERNET KIOSK IN THE LOBBY OF THE CREDIT UNION. THIS KIOSK ALLOWS MEMBERS WALK-UP ACCESS TO THEIR ACCOUNTS.

Built from Employee Commitment

Originally, CTFCU was run on a shoestring budget. Charles Blake served as the first manager and treasurer. Other employees were primarily volunteers, including Lulu Hodge, who took on the responsibility of keeping the books for the credit union. Hodge did all of the bookkeeping at home on her back porch until 1968, when the credit union—with a total of 1,635 members—finally moved into a tiny room at the board of education.

Marilee Greene came to work for CTFCU in 1970. Filling a position vacated by an employee who left one day at lunch hour and never returned, Greene proved to be an extraordinary employee. She worked every position from teller to manager, until becoming CEO in 1985.

As CEO, Greene worked hard to keep the credit union on the cutting edge by adding innovative services such as home banking and shared branching. Her service to the credit union industry did not go unnoticed. In 1996, Greene received the Frank Moore Outstanding Credit Union Professional Award for her dedication to the credit union movement. After 29 years of dedicated service, Greene retired in 2000, leaving behind a legacy of quality service to CTFCU.

On April 1st, 2000, Lynn Huether took over as CEO for CTFCU. The board of directors is confident that Huether will provide the leadership required to keep the credit union on the cutting edge of technology while offering its members the financial services they deserve.

Members Helping Members

Essential to the CTFCU philosophy is that the members themselves are in control of the credit union. Each credit union member votes to elect the officials who operate the organization. At CTFCU, the elected board of directors consists of fellow volunteer CTFCU members.

CTFCU offers a wide range of financial services, including checking and savings accounts, direct

deposit, debit/credit cards, and the Express Line phone automated teller. The credit union also provides a number of services tailored toward teachers. One of the special services available to teachers and school workers is the Summer Pay Plan. This account allows members to have a small percentage of their paycheck held in a special account that will help spread out their school year paychecks to cover the summer months.

With its Captain Cash Club, CTFCU makes it easy for children to learn the value of saving. The club has special prizes, mailings, and planned events designed to teach valuable lessons about the importance of a lifelong savings plan. Additionally, the Captain Cash Club is committed to educating children about other issues such as fire safety, the environment, and saying no to drugs. As an advocate of advanced education, CTFCU also offers a college scholarship program for its members up to age 23.

WWW.CLASSACT.ORG

The Internet has brought the opportunity for increased services to the credit union. With CyberBranch @Home, CTFCU members have 24-hour computer access to their accounts for inquiries, transfers, loan payments, and other services from work or home. All these transactions are performed over the Internet in real time so the member's account is automatically updated.

With the CTFCU Web page, information is available on estimated loan payments, current interest offered, and other financial information, as well as locations and office hours. These services are alternatives to traditional banking centers, and allow members to care for their financial needs at their own convenience.

PREPARED FOR THE FUTURE

CTFCU consistently strives to improve its services and convenience for its members. One step toward convenience was offering shared branching. The shared branch concept allows several credit unions to come together in one building, expanding the locations where members can access their accounts. The first shared branch opened in October 1999 on Dixie Highway. In addition, CTFCU members can do business nationwide through the Credit Union Service Corporation, the largest shared branch network in the United States.

CTFCU strives to offer progressive services to its members. Along with recent improvements in availability and Internet access, CTFCU recently finished the renovation of its main office, remodeling 8,450 square feet of office space and completing a 12,000-square-foot, two-story addition. One of the prime features of the new building is an Internet kiosk in the lobby of the credit union. This kiosk allows members walk-up access to their accounts. With touch screens to guide users through the process, this kiosk offers the latest in account availability to CTFCU members.

Built on a commitment to its members, Classroom Teachers Federal Credit Union plans to continue serving them with its innovative approaches.

WITH CYBERBRANCH@HOME, CTFCU MEMBERS HAVE 24-HOUR COMPUTER ACCESS TO THEIR ACCOUNTS FOR INQUIRIES, TRANSFERS, LOAN PAYMENTS, AND OTHER SERVICES FROM WORK OR HOME.

THE CTFCU BOARD OF DIRECTORS INCLUDES (FRONT ROW, FROM LEFT) MARY ANN BRUMAGEN; TOM MATHEWS; LYNN HUETHER, CEO; ROBERT RODOSKY; (BACK ROW, FROM LEFT) BEVERLY WESTPHAL; SAUNDRA PALMER; JO ANN ORICK; ELIZABETH CAPLES; SANDY WALKER-WOOLEY; AND IRVIN OWENS.

L&N Federal Credit Union

L&N Federal Credit Union traces its roots back to one of Louisville's oldest companies, the L&N Railroad (now CSX), which was originally named after its first run in 1859 from Louisville to Nashville. In the late 1800s and on through the first half of the 20th century, L&N Railroad was the premier rail line throughout the South. At the height of its day, L&N Railroad had more than 7,000 employees in Louisville alone. It was during this time of prosperity that the employees of the L&N South Louisville Shops began to explore the idea of establishing a credit union. In 1954, the credit union became a reality when 14 employees of L&N Railroad chartered the L&N Employees Federal Credit Union. For the next 28 years, the credit union served the L&N Railroad's employees and their family members exclusively, and experienced strong growth.

In 1982, the credit union expanded its field of membership through the adoption of select employee groups (SEGs). SEGs are businesses that would like to offer a credit union to their employees, but are too small to establish their own. To reflect the expanded field of membership, the credit union dropped "Employees" from the name, thus becoming the L&N Federal Credit Union.

Today, L&N Federal Credit Union is a thriving financial institution headquartered in Louisville, with eight other branches in Kentucky (four in Louisville, and one each in Corbin, London, Covington, and Florence). With more than 60,000 members and over 650 SEGs, L&N Federal Credit Union is the largest credit union in Louisville and ranks as one of the top in the nation.

WITH ROOTS IN ONE OF LOUISVILLE'S OLDEST COMPANIES, THE L&N RAILROAD, L&N FEDERAL CREDIT UNION TODAY IS A THRIVING FINANCIAL INSTITUTION. PICTURED HERE IS *The General* BY C.W. VITTITOW.

Complete Services

L&N Federal Credit Union is a not-for-profit financial cooperative, wholly owned by its members. Profits earned by the credit union are returned to its members in the form of higher returns on investments, lower interest rates on loans, and overall lower-cost services.

According to Callahan & Associates' 1998 Top 50 report, L&N Federal Credit Union ranked at the top of the list in two categories—total return to the member and return to the borrower. Both of these categories reflect the overall ability of a credit union to provide loans and services at the lowest possible cost.

From tiny, two-person companies to corporations with hundreds of employees, L&N Federal Credit Union can provide financial services to meet almost any need. Along with traditional financial services, L&N Federal Credit Union also offers investment options; mortgages and equity loans; 24-hour, automated telephone service; Internet home banking; indirect lending (financing of a vehicle with an L&N loan at the dealership); money market accounts; college scholarships; and more. L&N Federal Credit Union strives to be the

IN THE LATE 1990S, L&N FEDERAL CREDIT UNION BEGAN TO OFFER MORE HIGH-TECH SERVICES TO KEEP UP WITH ITS MEMBERS' NEEDS, SUCH AS INTERNET HOME BANKING. MONEYWEB®, THE INTERNET SERVICE, IS LOCATED AT WWW.LNFCU.COM AND ALLOWS MEMBERS FREE, WORLDWIDE ACCESS TO THEIR ACCOUNTS 24 HOURS A DAY, SEVEN DAYS A WEEK.

only financial resource its members will need.

Starting in the late 1990s, L&N Federal Credit Union began to offer more high-tech services to keep up with its members' needs, such as Internet home banking. MONEYWEB®, the Internet service, is located at www.lnfcu.com and allows members free, worldwide access to their accounts 24 hours a day, seven days a week. All transactions are performed in real time, giving members up-to-the-minute details on all accounts. For a small fee, members can make bill payments with the click of the mouse, rather than sending checks in the mail. Downloading information into Quicken and Microsoft Money is also available.

BUILDING RELATIONSHIPS

While putting an emphasis on technology and convenient services, L&N Federal Credit Union still places great importance on personal care. Following its mission—Building Permanent Relationships℠—the credit union lends a helping hand to those in need. When major flooding hit Louisville in 1997, many L&N members were affected. The credit union provided zero-interest loans to help those members recover from the disaster. The credit union's employees and their family members also have pitched in to help local charities over the years, including the American Heart Association and Kentucky Harvest, among others.

The service and concern provided for its members are typical of the L&N Federal Credit Union's ideals, and magnifies what Building Permanent Relationships is all about. As the company moves into the 21st century, L&N Federal Credit Union continues to work with the same values of trust, commitment, and family that have guided it since its inception.

SINCE ITS FOUNDING, L&N FEDERAL CREDIT UNION HAS BEEN BUILDING PERMANENT RELATIONSHIPS.

THOMAS INDUSTRIES INC.

While consumers might not be familiar with the name, chances are they have encountered some of Thomas Industries Inc.'s products. The company manufactures the compressors and vacuum pumps that comprise the heart of many different types of equipment. Its world headquarters in Louisville supports and manages eight production plants—five in the United States and three in Germany—and a sales organization that covers the globe.

Applications for Global Markets

Medical equipment, in particular, relies heavily on Thomas' products. In hospitals, laboratories, and medical and dental clinics around the world, Thomas' products can be found in all types of equipment designed to monitor, sterilize, diagnose, and analyze.

But Thomas' reach goes far beyond the medical field. The company's products are found inside many types of equipment that makes people's lives better and safer. "We specialize in being able to customize any compressor or vacuum pump that our customers need to fit their application," says Timothy Brown, chairman, president, and CEO.

Thomas' ability to offer innovative solutions and extensive knowledge to the original equipment manufacturer market is the foundation of its many long-standing customer relationships. For example, many different types of information technology equipment, including copy machines, printers, and photoprocessors, use Thomas compressors and vacuum pumps. "As the market demands smaller and smaller machines," says Brown, "Thomas has been able to fill the need with our miniature compressors, which allow all types of products to become smaller and more portable."

Today's automotive consumers may use a Thomas product in a variety of ways. The company's pumps provide safety and comfort to drivers and passengers in applications such as suspension systems and active seats, which reduce muscle stiffness for drivers of certain BMW models, for example.

Thomas' products are also hard at work making sure that special environmental equipment will keep people safe at work and at home. The company's compressors and vacuum pumps can be found in the equipment that treats water, instruments that monitor vehicle exhaust, and detection units that warn against leaks in underground fuel storage tanks.

A Significant Milestone

Thomas has been one of the world's leading manufacturers of fractional horsepower compressors and vacuum pumps since 1961. Until 1998, the company was also a leading manufacturer of lighting fixtures. That year, in an effort to focus its efforts on one specific product line, Thomas entered into a joint venture with The Genlyte Group, Incorporated to form Genlyte Thomas Group LLC. While the new entity manufactures lighting fixtures, Thomas now focuses on its core business—specialized compressors and vacuum pumps.

"Thomas Industries has been able to cover the globe and enter so many different markets because our goal has always been to seek out new markets and develop custom products for new as well as existing applications," says Brown. And as Thomas Industries Inc. sees it, global applications for its products are endless—virtually anywhere there is a need for pressure, vacuum, or flow.

Clockwise from top:

The vast majority of oxygen concentrators around the world contain Thomas pumps.

Thomas Industries Inc. is the global leader in the design and manufacture of compressors, vacuum pumps, and liquid pumps for a variety of applications.

Thomas' eight manufacturing facilities throughout the United States and Germany undergo continual improvement to maintain state-of-the-art machining, die casting, plastic injection molding, and robotics capabilities.

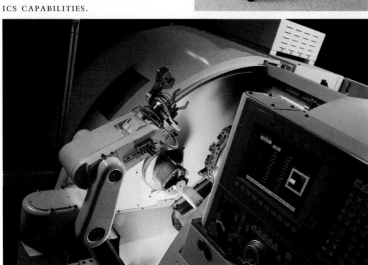

© DON SIVORI

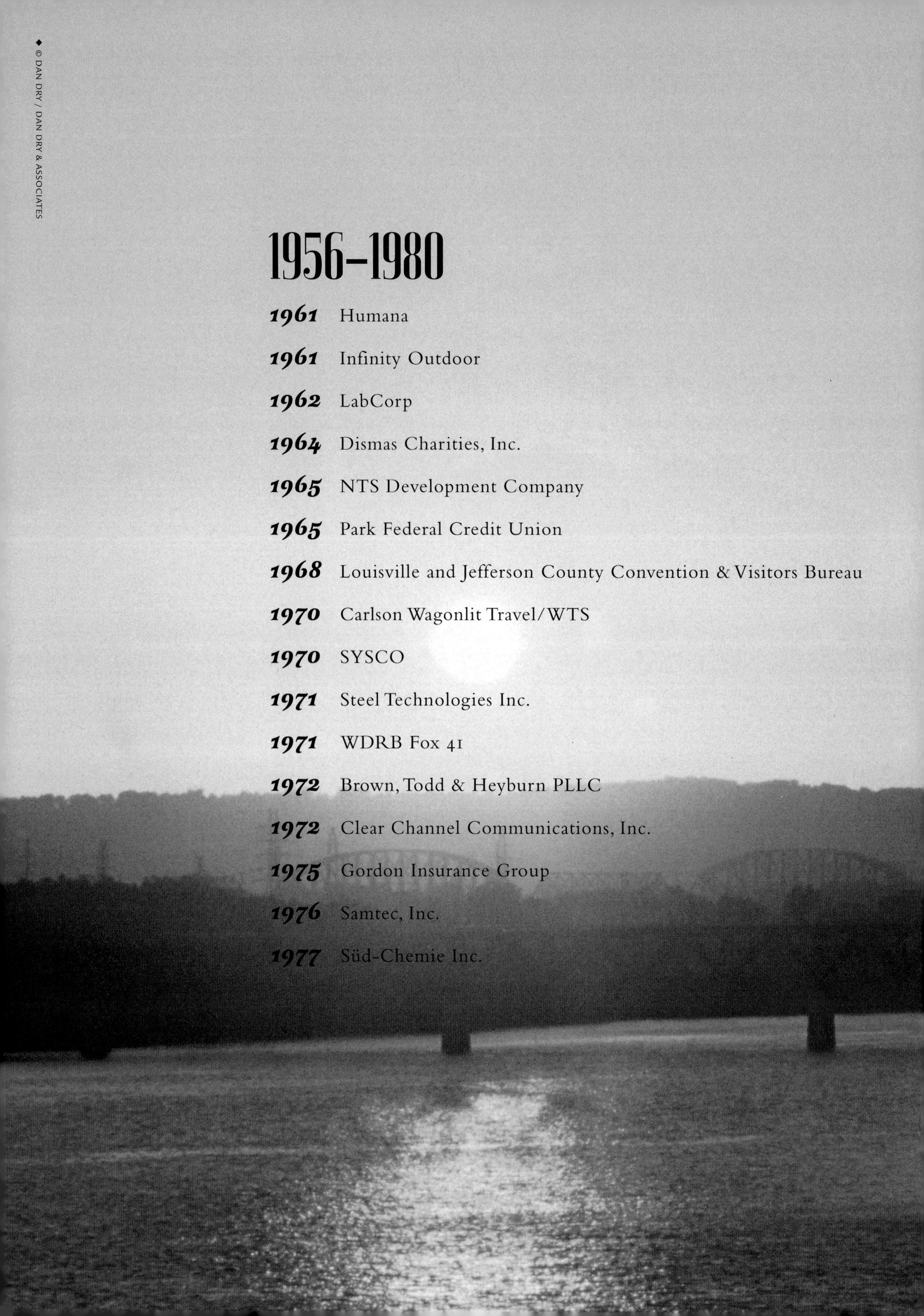

© DAN DRY / DAN DRY & ASSOCIATES

1956–1980

1961	Humana
1961	Infinity Outdoor
1962	LabCorp
1964	Dismas Charities, Inc.
1965	NTS Development Company
1965	Park Federal Credit Union
1968	Louisville and Jefferson County Convention & Visitors Bureau
1970	Carlson Wagonlit Travel/WTS
1970	SYSCO
1971	Steel Technologies Inc.
1971	WDRB Fox 41
1972	Brown, Todd & Heyburn PLLC
1972	Clear Channel Communications, Inc.
1975	Gordon Insurance Group
1976	Samtec, Inc.
1977	Süd-Chemie Inc.

Humana

A LEADER IN THE MANAGED CARE INDUSTRY, HUMANA IS REINVENTING itself to meet the health care needs of its members and affiliated physicians. This is not the first time Humana has gone through such a transformation. Humana was founded in 1961 by David A. Jones and Wendell Cherry as one nursing home with an initial investment of just $6,000, and eventually became the largest nursing home company in the United States. However, by 1972, Jones and Cherry had perceived the need for change; they reinvented Humana as a hospital company. Throughout the 1970s and 1980s, they focused on developing leading-edge hospitals with quality patient care.

Humana continued to create unique solutions to hospital industry problems. In 1984, Humana created Humana Health Care Plans, a family of flexible plans. In the early 1990s, Humana changed its focus again by separating its hospital and health insurance divisions, spinning off the hospital division as a new, separate company.

The Humana of today is a health care coverage organization—one of the nation's largest publicly traded managed health care companies—with 6 million medical customers located in 15 states and Puerto Rico. Humana offers coordinated health care coverage through a variety of plans, including health maintenance organizations, preferred provider organizations, point-of-service plans, and administrative services products.

Humana seeks to provide employers and individuals with simplified health insurance products that are affordable and easy to use. Humana's plans are available to employer groups, government-sponsored plans, and individuals. The company also offers an array of specialty products—such as dental, life, and disability insurance—that complement its core health offerings.

IMPROVING SERVICE THROUGH TECHNOLOGY

Humana is also focused on harnessing information technology to serve customers' needs better and more efficiently. Currently, Humana is transitioning to a paperless system where most member records will be maintained on an Internet-enabled computer.

"Humana's computer network will serve as the central hub where, with just the tap of a button, physicians' offices will be able to access patients' medical records, file insurance forms, and check on the status of claims," says Michael B. McCallister, Humana's president and chief executive officer. Others involved in health care paperwork, including office staff and insurance brokers, will also benefit from the time savings of this new system. And, improving the management of the company's technological resources makes handling member information easier and more efficient.

BOTTOM LEFT:
FOUNDED IN 1961 AS ONE NURSING HOME WITH AN INITIAL INVESTMENT OF JUST $6,000, HUMANA TODAY IS ONE OF THE NATION'S LARGEST PUBLICLY TRADED MANAGED HEALTH CARE COMPANIES, WITH 6 MILLION MEDICAL CUSTOMERS LOCATED IN 15 STATES AND PUERTO RICO.

BOTTOM RIGHT:
HUMANA IS LED BY MICHAEL B. MCCALLISTER (LEFT), PRESIDENT AND CEO, AND DAVID A. JONES, CHAIRMAN OF THE BOARD.

INNOVATIONS IN DISEASE MANAGEMENT

Humana's innovative disease management programs are moving the company into a new era in health care. Humana is a pioneer in the emerging field of disease management, which is comprised of various programs that identify patients with serious conditions and seek to provide these patients with better outcomes.

A crucial aspect of Humana's disease management program is the identification of patients with serious illnesses or conditions. Once such patients have been identified, their physicians and medical care facilities can provide more coordinated care. But disease management programs go a step further, working with the member directly to provide educational information, an overall treatment regimen, diet and nutrition information, and lifestyle change solutions.

An example of this innovation is the Cardiac Solutions disease management program for patients with coronary artery disease. Cardiac Solutions' specialized cardiac nurses work closely with plan members and their physicians to address clinical and behavioral issues such as smoking, high blood pressure, elevated lipid levels, lack of exercise, and poor medication compliance. Cardiac Solutions employs a comprehensive disease management system called MULTIFIT, which was developed at the Stanford University Department of Cardiac Rehabilitation and licensed exclusively to Cardiac Solutions. The company's progressive programs for disease management have received recognition from the Disease Management Purchasing Consortium and *Disease Management News* magazine.

HumanaBeginnings, an outgrowth of Humana's disease management programs, is a comprehensive prenatal program designed to support expectant mothers and try to improve pregnancy outcomes. The program seeks to help women have fewer low-weight births, fewer babies who require intensive care at birth, and better maternal health throughout pregnancy.

HumanaBeginnings works collaboratively with the patient and her physician to try to minimize risks. This program—which includes patient education materials such as a copy of the book *What to Expect When You're Expecting*—helps the patient and her physician by providing easy access to additional resources, including services such as home health care. The program also seeks to provide personal reinforcement for the patient and help in finding any community resources she might require.

Throughout the years, Humana has evolved with the vast changes in the health care industry to try to better serve its customers. But when asked to sum up the focus of his company, McCallister says, "While change will probably remain a constant for Humana, our desire to meet the needs of our customers swiftly and with responsibility will never change."

HUMANA'S MEDICARE PRODUCT DEVELOPMENT TEAM IS JUST ONE GROUP OF THE ORGANIZATION'S EMPLOYEES DEDICATED TO IMPROVING PATIENT CARE AND SERVICE.

Infinity Outdoor

MORE BUSINESSES THAN EVER BEFORE ARE SEARCHING FOR A VIABLE AND cost-effective means to attract a new audience. Competing with the multitude of television channels, Internet Web sites, radio stations, and newspapers, outdoor advertising is providing high value for the advertising dollar. Leading the way in this ever-changing industry is the Louisville firm Infinity Outdoor. ■ Infinity, a subsidiary of CBS Broadcasting, can trace its roots in Louisville to 1961, originally operating under the name Alwes Outdoor. In fewer than 40 years, Infinity has grown to become one of the premier outdoor advertising firms in the United States.

Today, Infinity operates the majority of the billboards in the Greater Louisville area. The company is the largest out-of-home advertising firm in North America, and operates outdoor bulletin, poster, mall, and transit advertising displays in more than 90 metropolitan markets in the United States, 13 metropolitan markets in Canada, and 44 metropolitan markets in Mexico.

INFINITY OUTDOOR'S MANAGEMENT TEAM INCLUDES (FROM LEFT) DON KING, MATT VERNON, KEVIN CARNES, MIKE SHEEHY, DIANE KAREM, AND CHRIS MASSLER.

THE FIRST FORM OF ADVERTISING

The history of billboards dates back to the 1800s, when circus acts and other traveling performers used bills posted on buildings to announce their performances. By the 1920s, with the birth of the national highway system, the idea of outdoor advertising had been transformed into a legitimate business.

In the modern era of outdoor advertising, tobacco and liquor ads accounted for a large percentage of billboard space. But as the industries and advertising regulations changed, more billboard space was used to advertise less traditional industries, in less traditional locations. By the end of the 1990s, everyone from health care companies seeking patients to Internet companies seeking visitors to their Web sites was using billboards to get messages out.

Of course, outdoor advertising has moved well beyond roadside billboards, finding its way into a multitude of locations in a variety of styles. Shopping malls, airports, stadiums, movie theaters, and supermarkets, as well as buses, subways, and trains, have all become bearers of outdoor advertising.

But perhaps the biggest change of all has resulted from technological advancements. Until recently, the majority of billboards were laboriously painted by hand. Today, thanks to significant technological advances, most billboards are printed on vinyl sheets using computer design and printing.

Applying preprinted billboards is far less time consuming than painting them by hand, providing a quicker turnaround for the client and a more cost-effective advertisement. With technological advances, the turnaround time for producing a billboard has gone from some 30 days to as few as five. Vinyl is also safer and quicker for the workers hanging the billboards, and can easily be taken down and stored for later use.

Enhanced designs on billboards also include the use of electronics for clocks, countdowns, and lighting effects. One of the most popular additions is the three-dimensional effects with objects extending over the top or side, or hanging below the billboard. Throughout this evolution, Infinity has maintained an industry lead in the design and implementation of these modern forms of advertising.

LOUISVILLE AND INFINITY: A PERFECT FIT

Through the years, Louisville has played a vital role in forging Infinity's success. The city, in conjunction with the vision of Infinity, has developed an excellent reputation as the top test market in the nation for innovative outdoor advertisements. As a midsize city where the majority of travel is by car on an elevated interstate system, Louisville provides its citizens with daily views of billboards and other types of outdoor advertising.

In its role as a test market, Louisville has provided several firsts to the outdoor advertising business. In 1972, Infinity became the first company to build billboards utilizing only a center pole. Until then, billboards were built with multiple poles or steel beams. The center-mounted billboards were easier to place, as well as more aesthetically

pleasing. Today, the mono-pole billboard is an industry standard.

Although the majority of Louisville-area billboards are computer printed on vinyl, the Louisville market has been known nationwide for the high-quality painted billboards lining its streets. Infinity painters in the Louisville market have been recognized as the top billboard painters in the country for some 20 years, according to *Sign of the Times,* a leading trade publication. Customers who are searching for a special advertisement and an individual statement often find that a hand-painted sign delivers a more vibrant and eye-catching statement.

As part of its commitment to the Louisville community, Infinity donates approximately 10 percent of its outdoor advertising space to a variety of charitable organizations. Groups such as Dare to Care, the American Red Cross, the Boys and Girls Clubs of America, and the Special Olympics have benefited from the quality advertising space provided by Infinity. From community commitment to quality advertisement, Infinity Outdoor continues to lead the way in its field.

INFINITY BILLBOARDS ARE USED BY A WIDE VARIETY OF LOUISVILLE BUSINESSES.

ONE OF THE MOST POPULAR ADDITIONS TO MODERN OUTDOOR ADVERTISING IS THREE-DIMENSIONAL EFFECTS WITH OBJECTS EXTENDING OVER THE TOP OR SIDE, OR HANGING BELOW THE BILLBOARD. THROUGHOUT THIS EVOLUTION, INFINITY HAS MAINTAINED AN INDUSTRY LEAD IN THE DESIGN AND IMPLEMENTATION OF THESE MODERN FORMS OF ADVERTISING.

LabCorp

The clinical laboratory is an essential component of modern, high-tech health care. As one of the nation's largest clinical laboratories, LabCorp is exceptionally qualified to serve physicians, hospitals, managed care systems, and occupational testing. Each day, procedures are performed in LabCorp's facilities on specimens from approximately 250,000 patients.

From Humble Beginnings to an Industry Leader

LabCorp plays a vital role as a key diagnostic service provider in the Louisville medical community. The company has been a part of Louisville since 1962, when it opened its first testing facility—located in the basement of what was then called Medical Towers North. Mary Ellen Inman, operations manager, has worked for LabCorp since those first days. "I have seen both our company grow and the testing itself expand dramatically. We've gone from doing standard, manual blood procedures to very complex, esoteric DNA testing with highly sophisticated equipment," says Inman. Today, the Louisville region of LabCorp does testing on more than 15,000 patients each day.

Nationwide, LabCorp is an innovative industry leader in the development and introduction of new clinical assays. Research, development, and technical support teams include a highly qualified group of MD- and PhD-level specialists, keeping LabCorp at the forefront of laboratory science. These laboratory experts are supported by 18,000 employees across the country.

Testing Services

In addition to standard testing, LabCorp also provides detailed testing for a variety of specific conditions. The company offers extensive diagnostic testing for oncology and cardiovascular assessment, as well as a comprehensive allergy assessment program. Also, the company offers a comprehensive prenatal assessment and an approach to reproductive testing that includes a thorough assessment of both parents. Physicians and patients can also rely on highly sensitive DNA testing for human immunodeficiency virus (HIV). "Our mission is to deliver quality information quickly, so our customers can focus on patient care and practice management," says Woodrow Cook, senior vice president of LabCorp.

At its Center for Esoteric Testing in Burlington, North Carolina, and its Center for Molecular Biology and Pathology in Research Triangle Park, North Carolina, LabCorp provides molecular diagnostics and other cutting-edge diagnostic technologies for physicians who require specialized, esoteric testing services. Methodologies employing molecular diagnostics often provide the most sophisticated or sensitive testing available, and are used for specialized oncology

LabCorp's Center for Molecular Biology and Pathology in Research Triangle Park, North Carolina, performs a variety of testing services, including polymerase chain reaction (PCR) testing.

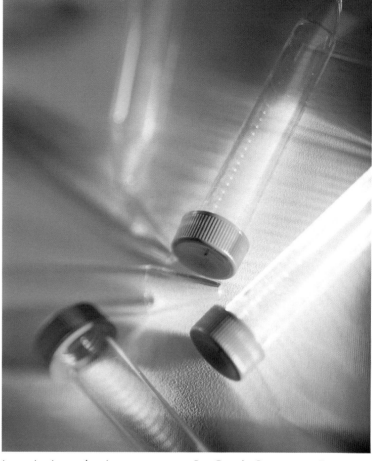

LabCorp's Center for Esoteric Testing is located in Burlington, North Carolina.

assays, genetics tests, infectious disease tests, and identity testing for paternity and specimen identification and matching.

LabCorp understands the sensitive nature of its work, striving diligently to offer excellence in clinical laboratory services. "Reliable results serve as the foundation of our corporate commitment to quality," says Cook. "Through total quality management, we seek to deliver our services more efficiently and consistently."

The quality assurance program developed by LabCorp goes well beyond minimum accreditation standards. Above and beyond inspections by state, federal, and other accreditation organizations, daily monitoring of results, internal proficiency challenges, and external programs serve as established protocols.

The company strives to provide personal and efficient service both to the medical community and to the patients it serves. By providing a national electronic data communication network, extensive courier routes, and strategic placement of facilities within close proximity to clients, LabCorp delivers effective and dependable daily service. Additionally, LabCorp provides client care coordinators who are available to assist customers by phone or in person with any questions regarding the company's laboratory services.

A Focus on People

People are at the heart of LabCorp's quality testing and superior customer service. From couriers and billing clerks to technologists and doctors, every LabCorp employee is focused on quality and customer satisfaction. Dedication to quality is also evident in the large number of long-term LabCorp employees: it is not uncommon to find 20- and 30-year employees in many key positions.

Ensuring that business is conducted according to the highest standards of ethical conduct and integrity is another important component of service for LabCorp. To achieve this goal, the company has established a compliance program to monitor conformity to federal, state, and local laws.

Even as the company strives to be a leader in its industry, LabCorp remains focused on the most important part of its business equation: the patient. "We always remember that at LabCorp we are working with people," says Cook. "From the technician who retrieves samples to the research analyst in the lab, each person is focused on the patient who relies on our results." LabCorp is dedicated to ensuring results that are consistently accurate and precise. This dedication to quality is essential to health care providers who rely on LabCorp's results when evaluating, diagnosing, treating, and monitoring their patients' conditions.

On the threshold of the 21st century, LabCorp stands as one of the finest and most extensive laboratory testing service providers in the United States. Constantly focused on its goal, LabCorp will continue to lead its industry for decades to come.

Dismas Charities, Inc.

THE 1960S PROVED TO BE ONE OF THE MORE TUMULTUOUS PERIODS in American history, and for Louisville, times were no different. But, as much of the world focused on racial issues and world politics, one man in Louisville was addressing the needs of a group less likely to generate much sympathy or support—those with criminal records. Through his efforts was born Dismas Charities, Inc., an organization that has grown to become one of the nation's largest not-for-profit providers of social services.

Today, the healing hand of Dismas extends far beyond Louisville, with facilities throughout the nation. Likewise, the scope of the organization's work has evolved beyond corrections. Dismas takes an active role in alcohol and drug treatment, job training, homelessness, domestic violence intervention, early childhood development, mentoring at-risk adolescents, and a vast array of other programs and services.

An Unpopular Crusade

During the early 1960s, the Reverend William Diersen, a Roman Catholic chaplain, ministered to the inmates of the Kentucky State Reformatory near LaGrange. During this time, Diersen was frustrated by the "revolving door," where people returned to prison for violating conditions of their release. He understood—like few others at the time—that without the necessary survival skills and support structure, such individuals tended to revert to a life of crime and would inevitably be returned to prisons, only to begin the cycle again. Diersen was determined to make a difference, and came up with the idea of a halfway setting where a person leaving incarceration could move into a supervised living situation that provided opportunities for independent decision making and accountability. The halfway facilities would help offenders in finding employment, reestablishing family and community ties, and managing tension and anxiety during the readjustment process.

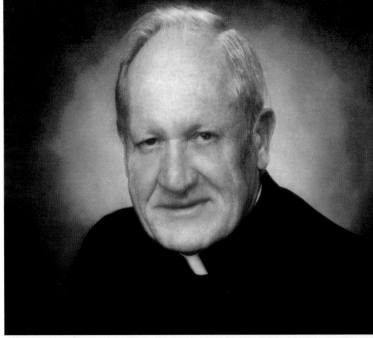

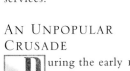

Reverend William Diersen is the founder of Dismas Charities, Inc. (top).

Dismas Charities Louisville is the oldest facility in the Dismas organization (bottom).

JOHN LAIR

Although Diersen knew he was on an unpopular crusade, he began seeking funds in Louisville to support his idea. Along with John Cannon, Kentucky's commissioner of corrections, Diersen sought and received sponsorship and support from the five councils of the local Knights of Columbus. "That sponsorship allowed Father Diersen to open the first Dismas facility," says Raymond J. Weis, president and CEO of Dismas Charities. "It was a 15-bed community corrections facility—or halfway house as it was then known—and was staffed with volunteers."

Dismas, named for the repentant thief who was crucified with Christ, was incorporated in August 1964, and the first facility opened in 1965. Members of the Knights of Colum-

bus continue to serve on the Dismas Board of Directors, the organization's official governing body.

Healing the Human Spirit

You'd be surprised some time if you reach down, give a helping hand, and maybe ask: 'How'd you fall? How can I help? Well, let's try this again.'" These words from Diersen sum up the mission of Dismas Charities, whose primary goal is healing the human spirit.

"Instead of just throwing people into jail or prison and later putting them back into society, we offer different, individually tailored programs that enable people to be more complete, more restored," says Weis. "Traditionally, prisons just house people and then return them to their communities after a certain number of years. But Dismas strives to work with our residents and prepare them to assume a responsible role when they are returned to their families and their communities."

As it has from the beginning, the single, most outstanding element in the operation of Dismas Charities remains its concern with the whole person. The organization addresses the factors that often lead to criminal behavior: poverty, alcohol, drugs, mental illness, lack of education, poor self-esteem, or even poor choices of peers and significant relationships.

"Programs are built around a holistic concept that seeks to address the emotional, spiritual, and physical aspects of our clients," says Jan Kempf, Dismas' executive vice president. "Along with trained, professional staff members, our clients set goals in each of these areas. This structure, with individual accountability, provides the framework for making appropriate choices."

"Dismas is about values," adds Weis. "We extend our hand to reach out and help somebody. We are driven by the need to assist people in changing, as well as by our desire to raise the community's consciousness as to how we should be dealing with those who have transgressed society. We do not endorse criminal behavior. What we do endorse is changing behavior so that we don't create more victims or destroy more families."

Part of the restorative justice provided by Dismas focuses on another often forgotten person in this chain of events—the victim. "Part of rehabilitating someone is to make them responsible for paying society back for their mistakes," explains Weis. "This is not a free ride. Our residents work, and part of their salary goes to pay back their victims."

Residents work in two ways. Through their paying jobs they pay restitution, fines, taxes, and child support, when warranted. In addition, they are required to pay 25 percent of their salary to the government to offset their confinement. They also provide services to the community at large through various community work projects.

An important key to Dismas Charities' high achievement level—the organization's model has a success rate of 85 percent—is

CLOCKWISE FROM TOP LEFT: THE DISMAS CHARITIES CORPORATE SERVICE CENTER IS THE LOUISVILLE-BASED HEADQUARTERS FOR THE ORGANIZATION.

THE CARRIGAN IS DISMAS' TRAINING CENTER.

DISMAS CHARITIES ST. PATRICK HOUSES STATE RESIDENTS IN A RENOVATED CIVIL WAR-ERA CHURCH.

PORTLAND CHILD DEVELOPMENT CENTER SERVES AS A DAY CARE AND LEARNING CENTER FOR AREA CHILDREN (LEFT).

KID'S CAFE PROVIDES HOT, NUTRITIOUS MEALS TO AT-RISK CHILDREN (RIGHT).

THIS STAINED-GLASS WINDOW SERVES AS DISMAS' CORPORATE LOGO (LEFT).

DISMAS' ST. ANN CAMPUS HOUSES 60 MEN, AND FOCUSES THEIR STUDIES ON SUCH HORTICULTURE PROJECTS AS THE MEDITATION GARDEN LOCATED ON THE CAMPUS (RIGHT).

addressing the family problems related to being incarcerated. "We focus on helping people make better choices, on how to effectively parent, and how to help the families learn how to reintegrate when one member has been absent due to incarceration," says Kempf.

In addition, the group's accomplishments stem from helping the residents find and retain jobs, and from involving the residents in community service programs. In Louisville, Dismas residents assist many community groups, including such services as maintenance of the Papa John's Cardinal Stadium, maintenance of local cemeteries, stocking local food banks, working at the Louisville Zoo and Humane Society, and a myriad of beautification projects. Since 1988, Dismas Charities residents have performed more than 1.5 million hours in community service. Based on today's minimum wage scale, that computes to more than $7.7 million in in-kind contributions to the nation's communities.

ONE MAN, ONE VISION

For 36 years, Dismas Charities has been a valued citizen and community partner to businesses and residents in neighborhoods where its facilities are located. "We try to work directly with the communities where we have facilities located and assess that community's need," says Weis. "Wherever there is a need, we try to fill it." So far, Dismas has responded by offering programs like Kid's Cafe to provide hot, nutritious meals to at-risk children, and facilities such as the Portland Child Development Center, which serves as a day care and learning center for area children.

Even the Dismas buildings themselves serve as solid examples of the group's connection to the community. "We frequently purchase vacant or unkempt buildings and, as we do with our clients, we provide attention, caring, and respect to their restoration," says Kempf. "Since 1964, Dismas has invested nearly $13 million in property and improvements in neighborhoods across Kentucky, Georgia, Florida, Texas, New Mexico, and Arizona. For example, Dismas headquarters is located in the old Joseph Seagram's Distillery administrative building on Seventh Street. Since purchasing it in 1998, the organization has refurbished the entire building and opened it to the community for a wide array of public events such as Breakfast with Santa and Derby Gala Ball."

From one man with a vision, one building, and a volunteer staff, Dismas Charities, Inc. continues to be a pioneer in addressing social issues and meeting the needs of those affected. Through its 35-plus years of service, the company has grown from that small but determined beginning to become one of the nation's top providers in community corrections. From its original 15 residents, Dismas has now served more than 75,000 individuals. Whether through community corrections, alternatives to incarceration, drug and alcohol treatment, parenting, mentoring, or other areas of its myriad services, Dismas continues to build on the ideas and vision set forth by its founder.

Says Weis, "We will continue our mission, Healing the Human Spirit."

LOUISVILLE AND JEFFERSON COUNTY CONVENTION & VISITORS BUREAU

Most people would probably be surprised to hear Louisville is the birthplace of the first electric trolley. It's also the home of the first cheeseburger, not to mention the Kentucky Derby and the Louisville Slugger baseball bat. Of course, Louisville is more than a home to these American icons; it is also a world-class tourist and convention destination, catering to more than 3.5 million visitors each year.

Founded in 1968, the Louisville and Jefferson County Convention & Visitors Bureau has only one goal—to promote Greater Louisville worldwide as the destination of choice for leisure and business travelers, trade shows, and conventions.

"The tourism industry is a very vital economic engine in our community," says Ronald L. Scott, bureau president and CEO. "And while the bureau is committed to the leadership position it enjoys, we are a membership organization, and the business and local community have always contributed to our success."

AN IDEAL CONVENTION LOCATION

Businesses and organizations choosing Louisville for meetings and conventions find a city with numerous first-class facilities and accommodations. "Louisville has many great things to offer the convention delegate," explains Susan McNeese Lynch, vice president of communications for the Louisville and Jefferson County Convention & Visitors Bureau. "The facilities we offer, our central location, the ease of transportation around our city, and our friendly population are all top rated." Louisville regularly hosts more than 400 conventions that combine to bring in nearly 700,000 delegates and generate more than $250 million in revenues annually.

Louisville constantly updates its facilities to ensure that it will continue to be a prime location for conventions and meetings. Indeed, the facilities in the community are outstanding. The Kentucky Fair and Exposition Center is the 10th-largest facility of its kind in the nation, with more than 1 million square feet of exposition floor space, and the newly renovated Kentucky International Convention Center offers nearly 300,000 square feet of premium downtown meeting space.

These excellent facilities have helped Louisville achieve a ranking as the fifth-best city for trade shows, allowing it to compete with larger cities such as Atlanta and Las Vegas. Currently, Louisville is home to four of the top 20 trade shows in the nation.

NOT JUST FOR BUSINESS

Travel and tourism revenue plays a very important role in the Louisville economy. Lynch is quick to point out that the travel and tourism industry in Jefferson County, a $1.1 billion industry annually, is directly responsible for more than 26,000 jobs, and generates nearly $220 million in tax revenues for federal, state, and local governments.

Today, Louisville is attracting more leisure travel groups than ever before. Visitors to the city have more than 40 world-class events and festivals from which to choose, including the Kentucky Derby, Equitana USA, the World Championship Horse Show, and the Kentucky State Fair. When the events are over, Louisville has more than 80 attractions to offer, including the Louisville Slugger Museum, the Kentucky Derby Museum, and Six Flags Kentucky Kingdom.

What once was a sleepy little river town on the banks of the Ohio River is today a thriving metropolis, providing excellent resources and accommodations to the largest trade show or the smallest tour group alike. As the city continues to grow, the Louisville and Jefferson County Convention & Visitors Bureau will continue to lead the way in promoting the Greater Louisville area as an exciting travel destination.

VISITORS TO THE CITY HAVE MORE THAN 40 WORLD-CLASS EVENTS AND FESTIVALS FROM WHICH TO CHOOSE, INCLUDING THE KENTUCKY DERBY (BOTTOM RIGHT), EQUITANA USA, THE WORLD CHAMPIONSHIP HORSE SHOW, AND THE KENTUCKY STATE FAIR. WHEN THE EVENTS ARE OVER, LOUISVILLE HAS MORE THAN 80 ATTRACTIONS TO OFFER, INCLUDING THE LOUISVILLE SLUGGER MUSEUM (TOP), THE KENTUCKY DERBY MUSEUM, AND SIX FLAGS KENTUCKY KINGDOM (BOTTOM LEFT).

NTS Development Company

LOUISVILLE'S FOUNDING FATHERS BUILT THE CITY'S QUALITY OF LIFE brick by brick, constructing the kind of stately neighborhoods and vibrant commercial districts that made Louisville famous. Today, that spirit of excellence lives on in Louisville's newest fine homes and office complexes, thanks in large part to the efforts of NTS Development Company. ■ A drive through Louisville's most picturesque developments reveals the legacy of 35 years of NTS in the city: the custom-built estate homes of Lake Forest, Sutherland, and Oxmoor Woods; luxury apartment homes of the Willows; and booming commercial and industrial developments of Blankenbaker Business Center and Plainview Center, to name a few.

These are, however, just a fraction of NTS' total developments, which include some 6,700 custom homes, 3,500 luxury apartments, 2 million square feet of business center space, 1.5 million square feet of executive office space, 700,000 square feet of industrial space, and 456,000 square feet of retail space. Today, NTS has more than 722 acres of land under development, bringing the total development to more than 6,500 acres of land in 11 major markets.

NTS is, in fact, a diversified, full-service real estate company that provides property management, construction development, and asset management for its wholly owned properties as well as institutional establishments. No matter what the scope of the project, every NTS property has one thing in common: excellence—the sort of excellence that comes from only the very best materials, workmanship, and management.

SOME OF NTS DEVELOPMENT COMPANY'S NOTABLE PROJECTS INCLUDE LAKE FOREST (TOP), BLANKENBAKER BUSINESS CENTER (MIDDLE), THE WILLOWS APARTMENT HOMES (BOTTOM), AND ANTHEM CENTER (OPPOSITE).

J.D. Nichols, NTS chairman and CEO since 1965, puts it simply: "From the very beginning, we decided we'd try to do things differently—to set a higher standard than the typical development. Fortunately, we found there was a market for high-profile, luxury developments, and it has been our driver ever since. This may sound simplistic, but we always try to do just a little better with each new project we start. That way, you're always topping yourself—and your customers' expectations."

Turning Houses into Communities

While the company had developed several apartment communities, as well as LaFontenay Apartments in the East End, in the 1970s, its first major endeavor was Plainview, now one of Louisville's most desirable communities and an anchor for the bustling Hurstbourne commercial corridor. Plainview was a carefully planned, environmentally sensitive development that offered a unique, contiguous neighborhood encompassing a residential community including multifamily, commercial, and retail areas.

The success of Plainview propelled NTS into rapid growth and the founding of Lake Forest in 1983. Situated on 1,100 acres of rolling meadows and shady woodlands, the community includes 30 lakes and more than 1,800 homesites built by quality Louisville builders. Lake Forest is also home to Lake Forest Country Club, an Arnold Palmer-designed, 18-hole championship golf course.

"Lake Forest represents everything we want to achieve in an NTS community," says Nichols. "It's for the kind of person who wants to live life to the fullest and will settle for nothing less than the best. I think it's safe to say that the only communities that can match Lake Forest's reputation might be another NTS development, like Sutherland or Oxmoor Woods."

Building from its early success with Plainview and Lake Forest, NTS has developed numerous new communities in Louisville and Lexington, eventually expanding to Indiana, Virginia, Georgia, and Florida, although its headquarters remains in Kentucky.

The hallmark of every NTS residential development is preservation of the landscape's natural beauty and a desire to create a true community life for its residents, including amenities such as a community lodge, swimming pools, jogging and bicycle trails, picnic facilities, playgrounds, and soccer fields. Currently, the company has communities under development including Lake Forest, Sutherland, and Glenmary, all in Louisville; Fawn Lake in Fredericksburg; and Lake Forest of Orlando. Previously developed communities in Louisville include Plainview, Owl Creek, Oxmoor Woods, the Springs, Stone Bridge at Anchorage, Copperfield, English Station, and Douglass Hills Estates.

NTS applies the same standard of excellence found in its single-family residential communities to its luxury apartment communities, featuring lush landscaping, cobblestone walkways, marble foyers, spacious closets, master baths, clubhouses, and athletic facilities. The company owns and manages more than 3,500 existing luxury apartments in Kentucky, Indiana, and Florida.

Building the Commercial Properties That Build Business

The pace of change in American business has brought about the need for more services and amenities for businesses than ever before. It's a challenge Nichols is glad to meet, having built and developed more than 6 million square feet of commercial business space.

"Just providing a shell of an office space isn't good enough for today's businesses," Nichols says. "They need convenient locations, consolidated operations, and prestigious surroundings. They also need a very flexible work space. We work hard to give them the options they need."

NTS' clients have their choice of executive office buildings, such as the Atrium Center, Springs Office Center, and Plainview Center, or multiuse office service centers, such as Blankenbaker Business Center and Commonwealth Business Center, which allow businesses to consolidate warehousing, manufacturing, and headquarters operations in one complex.

For the discriminating customer needing superior office space, NTS' executive office buildings offer top-of-the-line office space with spacious private parking lots, manicured landscaping, and access to high-tech telecom services.

Then there are the companies who choose to design their own executive office building. For them, there are developments like Blankenbaker Crossings, a 550-acre commercial development that allows manufacturing facilities to blend effortlessly with their headquarters offices. Owners of these properties enjoy easy access for tractor trailers; expansion space for warehousing, distribution, and light manufacturing; and the electronic access and maintenance needed for headquarters offices and showrooms.

Total Turnkey Management of Commercial Property

In addition to managing its own holdings, NTS has a staff of more than 300 property management experts who also work to manage the properties of other owners.

One of NTS' services is construction management, where NTS can bring its decades of building experience to the table for its clients. In these cases, NTS handles every aspect of the project, including hiring the general contractor, obtaining permits, working with architects, and finally, leasing and managing the finished building. Owners enjoy total peace of mind and a solid return on their investment.

As part of this initiative and many others, NTS' president, Brian F. Lavin, is leading the company in providing services for institutional investors. "With our long experience, we can now offer diversified, full real estate services for property owners," Lavin says. "We know the business, and we have a proven track record of success. Owners can hire us with confidence and know we'll get the job done right, and done to NTS' high standards for excellence."

A Commitment to the Community

Ultimately, NTS is about more than just bricks and mortar, marble fireplaces, and swimming pools. It's about building the infrastructure and essence of a community.

"It is very, very important to us to become one with the land and the community," Lavin says. Every NTS property is developed in close coordination with government agencies, such as planning, zoning, and beautification leagues. "It's just good business," says Lavin. "If you work smart and preserve what is valuable and special about a place, it only increases the value of the final project. Our reputation for responsible development makes it very easy to get things done."

NTS' employees show their commitment to the community in other ways as well, donating their time and effort to many Louisville-area charities, ranging from the National Association of Home Builders to the Olmsted Parks Conservancy, Fund for the Arts, Metro United Way, and March of Dimes, to name just a few. Nichols serves on the board of the Louisville Regional Airport Authority and has been heavily involved in its recent expansions.

Lavin has a prediction for the future of NTS in the community: "We feel that if we stick to our core philosophy—creating quality properties and delivering excellent value for our investors—we'll continue to be successful. NTS is part of the fabric of the community now, and we're very excited about our future."

Park Federal Credit Union

Since its founding in 1965, Park Federal Credit Union's mission—to serve the employees of General Electric's (GE) Appliance Park of Louisville—has expanded to include other GE facilities, along with numerous select employee groups. With branches in Alabama, Indiana, and Texas, as well as holdings in Louisville, Lexington, Richmond, and Florence, Kentucky, Park Federal now serves more than 50,000 members at more than 500 companies.

A member-owned, not-for-profit financial cooperative, Park Federal Credit Union provides a full range of quality financial services at competitive prices, and is led by a volunteer board of directors, a three-person supervisory committee, and a staff of more than 125 employees. Park Federal Credit Union encourages ongoing interaction between the staff and the board of directors, who are active members participating in numerous committees to manage the policies and planning of the credit union.

Credit Union Service Philosophy

The credit union system has worked successfully in this country for more than 85 years, and the movement has grown and expanded greatly during the 1990s. During that time, credit unions rated number one among financial institutions in customer satisfaction, according to *American Banker*'s annual customer satisfaction survey. Nationwide, more and more people join credit unions every year; in fact, today, there are more than 12,000 credit unions, which serve more than 70 million people.

The driving force behind the movement is the central core of the credit union philosophy, which states that a credit union exists not for profit, not for charity, but for service to its members. Credit unions are organized to serve and to provide their members with a safe place to save and borrow money at reasonable rates.

In effect, members pool their funds to make loans to one another. In order to join a credit union, individuals must first be eligible for membership. At Park Federal Credit Union, employees of participating companies and their immediate family members—including those living in the household, as well as siblings, parents, and grandparents—are eligible for membership.

"There are many benefits that come from being involved in the credit union movement," says Gary McClure, Park Federal's Business Development Manager, "not the least of which are the generally lower rates we charge our members for loans, and the generally higher rates we pay our members for savings."

Hassle-Free Employee Benefits

As the credit union movement continues to expand, Park Federal Credit Union is actively pursuing companies to become a part of its financial cooperative. The benefits of providing credit union membership translate into an attractive addition to the company's benefits package. While credit union access is limited to sponsor companies, any company can apply to become a sponsor, final approval for which is determined

Park Federal Credit Union offers a wide range of financial services designed to help members reach their financial goals.

PARK FEDERAL CREDIT UNION HAS NINE CONVENIENT LOCATIONS, INCLUDING THREE IN GREATER LOUISVILLE.

by the National Credit Union Administration.

Park Federal Credit Union accepts applications from interested companies, and after an application is approved, the credit union takes care of all administrative responsibilities. To ensure that new members receive all of the necessary information, Park Federal Credit Union hosts on-site information seminars or displays, as well as membership awareness programs.

A VARIETY OF SERVICES

Park Federal Credit Union offers a wide range of services, extending from basic savings and various consumer loans to investment services and real estate loans. The credit union offers ATM, debit, and credit cards, as well as free, 24-hour telephone access and electronic bill payment. In addition, its checking accounts do not have any service fees, allowing a typical member to save a significant amount of money each year on this fee alone. Making automobile loans more available, Park Federal Credit Union is involved in the DealerDirect program, which allows members to close loans at participating dealerships.

Park Federal Credit Union's InTouch@Home service gives the credit union's members access to their accounts 24 hours a day via the Internet. Members can check their balances, transfer funds between accounts, make loan payments, and print out statements when it is most convenient for them. Using the Internet to its full advantage, Park Federal Credit Union now holds regular on-line chats at its Web site on subjects such as individual retirement accounts and investment services.

PEOPLE HELPING PEOPLE

Service to its members is not the only way Park Federal Credit Union contributes to the community. Members and staff have contributed time and financial resources to programs that assist the less fortunate. In recent years, Park Federal Credit Union has repeatedly earned the Kentucky Credit Union League's Dora Maxwell Award for Outstanding Social Responsibility (a state winner in eight of the last 10 years), and in 1996, placed second in the same national award among credit unions of similar asset size.

"Over the years, we have made certain that our members come first," says Ken Mattingly, president, "and Park Federal will contain our purpose, which is to support the philosophy of people helping people."

MEMBERS OF ALL AGES FIND PLENTY TO SMILE ABOUT AT PARK FEDERAL CREDIT UNION.

Carlson Wagonlit Travel/WTS

Whether a traveler is going to Chicago on business, to Texas for a convention, or off to a romantic Caribbean hot spot, Carlson Wagonlit Travel/WTS can help with all of the necessary details. For more than three decades, the company has been providing comprehensive corporate and leisure travel planning, as well as a variety of other travel-related services. Although its name has changed over time, the firm has remained under the steady leadership of Chairman of the Board and CEO Joseph Ferguson since its founding in 1970.

"Originally, we started out as an independent travel agency, but in an attempt to expand our services, we became associated with the Woodside Travel consortium," Ferguson says. The company was known locally as Woodside Travel for more than 20 years, and continued its growth by acquiring smaller independent travel agencies throughout Kentucky, Tennessee, Ohio, and Indiana.

In 1995, to broaden services even more, the company became associated with one of the largest travel management companies in the world—the Carlson Wagonlit Travel group—and changed its name to Carlson Wagonlit Travel/WTS. As a wholly owned and separate company, it is the largest associate in the Carlson Wagonlit Travel group, which has more than 4,000 national and international offices. Through the dual approach of joining large travel associations and acquiring quality independent travel companies, Carlson Wagonlit Travel/WTS has become one of the top travel companies in the United States and rates as the 38th-largest travel agency in the nation.

Travel Management for Business

Carlson Wagonlit Travel/WTS is probably best known for quality and comprehensive travel management services for business. "When I started in the travel business 30 years ago, nobody was providing the services that corporations needed to assist the business traveler," says Ferguson. "I knew there was a niche in the business for somebody who could help corporations actually manage their travel and help them save money."

Instead of just making travel arrangements, Carlson Wagonlit Travel/WTS provides detailed corporate management reports that can help companies determine how much they are actually spending on travel. "We can help the company with everything from negotiating rate reductions with frequently used hotels to ensuring that their employees are following company travel policies," says Ferguson.

Other important services that Carlson Wagonlit Travel/WTS has carved out just for business include a 24-hour customer service line that is open 365 days a year. The company handles everything from booking arrangements to travel emergencies. And Carlson Wagonlit Travel/WTS uses the most modern computerized systems to ensure that travelers always have access to the lowest rates available. Carlson's proprietary, state-of-the-art quality assurance system not only continuously seeks a lower fare, but verifies that there are no inconsistencies in the itinerary, a needed rental car will be at the correct airport at the correct date and time, and the requested hotel room has been booked correctly.

The New Albany, Indiana, retail outlet of Carlson Wagonlit Travel/WTS is one of 27 travel stores located in Indiana, Kentucky, Tennessee, and Ohio.

Carlson Wagonlit Travel/WTS' executive team includes (standing, from left) Caroline Donnelly, executive vice president and COO; Tom Lumley, president; JoAnn Clark, vice president leisure; and (seated) Joseph Ferguson, chairman and CEO.

Business travelers at the Louisville International and Lexington Bluegrass airports have access to Carlson Wagonlit Travel/WTS business centers, which include access to private workstations, fax machines, copiers, telephones, and, of course, travel agents.

MEETINGS, CONVENTIONS, AND GROUP TRAVEL

Carlson Wagonlit Travel/WTS works with its clients and ad hoc groups to provide group travel services. Qualified travel professionals at Carlson Wagonlit Travel/WTS will provide recommendations and alternative destinations consistent with budget requirements. This service is available for as few as 20 participants, although most groups can be significantly larger. These services are all-inclusive, including ground transportation, site inspections, and tours of local interest. In most cases, a negotiated hotel and airline contract will significantly reduce the group's costs.

As a destination management company, Carlson Wagonlit Travel/WTS assists those companies who are inbound to the Louisville area with particular expertise in local trips and excursions.

LEISURE TRAVELERS

Although Carlson Wagonlit Travel/WTS has been known as a premier business travel management company for 30 years, it is equally recognized as being a travel service agency with high-quality programs for the leisure traveler.

Combining a local and national presence, Carlson Wagonlit Travel/WTS uses its buying power to obtain better and more economical travel opportunities for its clients. This means that travelers with only limited time, as well as those who plan a more extended vacation, can be assured that they will receive the best rates and the most reliable services available.

CONTINUED GROWTH

Carlson Wagonlit Travel/WTS continues to grow and diversify to meet the needs and demands of the traveling public. The company recently acquired TriState Travel School, an accredited provider of up to 300 graduates a year in the area of travel and hospitality services. This school allows CarlsonWagonlit Travel/WTS to provide the best-trained professionals in the industry.

"We really intend to be the one travel company that offers everything the traveler needs. We provide complete travel management for businesses, group travel and meetings, and a wide variety of leisure travel. And now with such a large travel school, we can ensure that our customers will have the best travel consultants in the industry," says Ferguson.

THE NEW CORPORATE RESERVATION CENTER SERVING SCOTTSBURG, INDIANA, IS ONE OF FIVE LOCATED THROUGHOUT THE REGION.

THE CARLSON WAGONLIT TRAVEL/WTS CORPORATE MANAGERS INCLUDE (FROM LEFT) PAM MITCHUM, CHERYL FERRILL, VICKI ZORE, RON BYRD, LISA BROOKS, LEANNE WOOSLEY, CINDY BOWLING, LINDA WYATT, SONJA MCALISTER, JIM RICHEY, AND BARBARA POWELL.

SYSCO

With sales of $19.8 billion, SYSCO is the largest food service marketing and distribution company in North America, serving about 325,000 restaurants, health care facilities, educational institutions, and other customers. The company operates from 78 locations throughout North America. ■ SYSCO's line of 275,000 products includes a broad assortment of fresh and frozen meats, seafood, poultry, fruits, vegetables, fresh produce, bakery goods, canned and dry foods, paper and disposable items, sanitation products, dairy foods, and beverages, as well as kitchen and tabletop equipment and medical and surgical supplies.

A Glance at the Past

SYSCO/Louisville Food Service Company plays a vital role as part of the nationwide group of SYSCO Food Service Companies. The creation of the branch dates back to 1901, when it was only a small retail grocery house called Louisville Grocery Company. James and William Glazebrook formed the company to deal in retail groceries, coffee roasting, and packaging.

In December 1901, the Louisville Grocery Company held the first meeting of its board of directors. One of the important decisions made at that meeting was to purchase two mules in good health and suitable for pulling a delivery wagon. This decision was the first of many that have demonstrated the dedication of the company to providing the best in service to its customers. It was not until after World War II that the company moved into the institutional field.

Frank M. Ellis purchased Louisville Grocery Company in 1961, and moved it to the Louisville Industrial Center in 1963. Throughout the years, a concern for the needs of its customers—combined with a careful attention to the quality of its products and a willingness to make innovative changes in answer to market needs—continued to drive the company's business philosophy. In 1969, John Baugh, Ellis, and seven other private owners merged their companies to form SYSCO (Systems and Services Company), forging the foundation for the most powerful food service distribution corporation in North America.

Today, SYSCO employs more than 600 people in Louisville and serves as a one-stop distributor offering a wide range of food products, supplies, and equipment. More than 11,500 items are delivered to customers in a 250-mile radius of Louisville.

Committed to the Community

SYSCO employees participate in several outreach programs for people in need. A supporter of the Metro United Way campaign, the company became partners with the American Cancer Society in its Making Strides Against Breast Cancer 5K Walk-A-Thon. SYSCO employees donated necessary items, collected pledges, and walked in memory of friends or loved ones.

With an intense desire to continually improve the quality of the area's schools, SYSCO employees donate their time to visiting middle-school students. Employees spend time individually with two or three students, discussing the importance of staying in school.

Natural disaster victims have also received assistance from SYSCO. The company donated paper products and cleaning supplies to victims of a 1998 tornado in Shepherdsville, Kentucky, and bottled water to flood victims in Des Moines.

Dedicated to Its Employees and Its Customers

SYSCO recently developed the CARES (Customers Are Really Everything to SYSCO) Program. In this program, the company strives to dedicate itself to the happiness and growth of its employees and customers. SYSCO feels that if employees are treated well, they will pass this philosophy on to their customers. "We embrace the CARES philosophy for both our employees and our customers," says President and CEO Peter J. Scatamacchia. "As SYSCO continues to grow, the dedication of our employees will allow us to continue to set high standards," Scatamacchia continues. "Their commitment helps to promote SYSCO as a company of choice in our marketplace."

Peter J. Scatamacchia is president and CEO of Sysco/Louisville Food Service Company.

BROWN, TODD & HEYBURN PLLC

Kentucky's largest law firm, Brown, Todd & Heyburn PLLC, has a practice with unlimited boundaries. Created by a 1972 merger of three Louisville firms, the original entity had only 20 attorneys. At the dawn of the new millennium, Brown, Todd & Heyburn has more than 225 professionals located in Kentucky, Indiana, and Tennessee.

DEDICATED TO EXCELLENT SERVICE

Since its inception, Brown, Todd & Heyburn has been dedicated to quality and outstanding client service. The firm's practice capabilities expand the full range of business law, real estate, and litigation.

Responsiveness and an excellent work product are the standards by which Brown, Todd measures itself. The firm strives to help clients successfully develop their businesses. Clients include emerging entrepreneurial businesses, as well as national and international Fortune 500 companies. "We remain dedicated to proven strategies on behalf of our clients while incorporating innovative approaches at every opportunity," says C. Edward Glasscock, managing partner of Brown, Todd. "Our goal has always been excellent service. As our client base continues to grow, so will our services, efforts, and resources."

Whether working in the digital marketplace with e-law and the growing area of cyberspace, or representing owners, trainers, and breeders of some of the world's finest Thoroughbreds, Brown, Todd & Heyburn is poised to advance the interests of its varied clients.

The firm's dedication to solving clients' changing legal needs and its focus-on-the-client-first attitude have not gone unnoticed. In 1993, *International Corporate Law* magazine named Brown, Todd one of the top 14 law firms in the country. Working with numerous foreign and domestic companies engaged in cross-border transactions and providing a leadership role in international trade assistance for this region, Brown, Todd is the only Kentucky law firm to receive the Kentucky World Trade Center's World Trade Success Award.

DEDICATED TO THE COMMUNITY

In addition to its groundbreaking work in the legal arena, the firm maintains a long-standing commitment to corporate citizenship, evidenced by its contributions of both human and financial resources to worthy causes in each city in which it has offices. Brown, Todd's lawyers serve on boards of many civic, charitable, and nonprofit organizations. Each office honors the firm's tradition of providing pro bono legal services for those in need. As the firm continues to evolve within the new economy, Brown, Todd's commitment to corporate citizenship remains central to its identity.

There is a saying that success is not a destination but a journey. As long as the journey continues, this law firm will be there. Brown, Todd & Heyburn truly embodies its premise: Your Success Is Our Success.

BROWN, TODD & HEYBURN PLLC IS LOCATED IN AEGON CENTER IN DOWNTOWN LOUISVILLE.

STEEL TECHNOLOGIES INC.

STEEL TECHNOLOGIES INC. IS RECOGNIZED AS ONE OF THE LEADERS in the steel processing industry. Headquartered in Louisville since 1971, Steel Technologies services a number of industries from 16 manufacturing operations located throughout the United States and Mexico. ■ Merwin J. Ray founded the company in 1971 in Eminence, Kentucky. The company was originally named Southern Strip Steel to reflect Ray's desire to be a southern-based manufacturer of processed strip steel products. Southern Strip Steel grew steadily, and by 1985, sales had reached $54 million. Ray then decided to take the company public and the name was changed to Steel Technologies. The company continued its dramatic growth initially by greenfield plant startups. Additional growth in recent years has been accomplished through selective acquisitions in similar steel processing businesses in new geographic markets, and 1999 was a record year in both sales and earnings. The company has continued to keep its sights set on providing service and quality to its customers, while creating an exciting culture and stimulating work environment for all of its employees with a dedicated commitment to rewarding its shareholders.

VALUE-ADDED SERVICES

Steel Technologies performs a number of high-value-added services in the intermediate steel processing industry. These services include pickling, slitting, rolling, cutting to length, blanking, leveling, edging, annealing, and custom fabricating, among others. These various manufacturing operations serve to convert flat rolled steel—produced from various steel mills—into usable products that the company's customers can stamp or process into manufactured components. These manufacturing operations fill a valuable niche between the primary producers of steel and the original equipment manufacturer's customers, while at the same time improving quality and service.

Increased quality demands in the end user steel community have paved the way for many growth opportunities for Steel Technologies. Beginning with flat rolled steel coils that weigh as much as 80,000 pounds and are 72 inches in width, Steel Technologies provides a final product to exact width, weight, surface, hardness, chemistry, and shape. "Achieving what the customer wants when they want it is why we are in business," says Michael J. Carroll, president and chief operating officer.

"Our processes ensure the quality of product, but our people ensure the utmost customer service," Bradford T. Ray, vice chairman and chief executive officer, points out. "Each of our specific strategic business units is located geographically close to our customers. Because our employees at these locations know their clients best, we let each specific strategic business unit service its own clients within its region, so the unit can really focus in on its clients and service. The corporate office does the centralized functions, but the hands-on customer service is done at the plant levels."

Steel Technologies successfully serves the steel supply requirements for a variety of industries. The company has about 700 customers in the automotive, lawn and garden, construction, appliance, agricultural, and office equipment industries. Each of these industries has its special demands, and Steel Technologies has received many quality and service awards, including QS 9000, Ford Q1, Chrysler PentaStar, GM Mark of Excellence, and more.

IN 1971, MERWIN J. RAY FOUNDED SOUTHERN STRIP STEEL, WHICH BECAME STEEL TECHNOLOGIES INC. IN 1985.

STEEL TECHNOLOGIES PERFORMS A NUMBER OF HIGH-VALUE-ADDED SERVICES IN THE INTERMEDIATE STEEL PROCESSING INDUSTRY. THESE SERVICES INCLUDE PICKLING, SLITTING, ROLLING, CUTTING TO LENGTH, BLANKING (PICTURED HERE AT THE EMINENCE, KENTUCKY, FACILITY), LEVELING, EDGING, ANNEALING, AND CUSTOM FABRICATING, AMONG OTHERS. THESE VARIOUS MANUFACTURING OPERATIONS SERVE TO CONVERT FLAT ROLLED STEEL—PRODUCED FROM VARIOUS STEEL MILLS—INTO USABLE PRODUCTS THAT THE COMPANY'S CUSTOMERS CAN STAMP OR PROCESS INTO MANUFACTURED COMPONENTS.

STEEL TECHNOLOGIES' CORPORATE HEADQUARTERS IS LOCATED IN LOUISVILLE.

DEDICATED TO THE COMMUNITY

The company employs more than 1,000 people nationwide and supplies steel for customers worldwide, according to Ray, but it is still not a household name in Louisville. "We are the best-kept secret in Louisville," says Ray. "Because we don't produce a retail product, people don't know too much about us. It is likely, though, that we had something to do with the steel in your car, your office furniture, or even your golf clubs."

Although Steel Technologies has a multinational presence, the company remains dedicated to the community. The corporate office in Louisville handles accounting, purchasing, and many customer service and human resource functions, as well as headquarters for the executive management team. There are nearly 100 employees based in Louisville, with the number expected to double by 2005. Having grown far beyond the first operation in Eminence, Steel Technologies looks for even greater growth in the future, while remaining firmly grounded in Louisville.

ANNEALING IS A VALUE-ADDED HEAT-TREATING PROCESS TO PREPARE STEEL FOR FURTHER PROCESSING. STEEL TECHNOLOGIES USES 100 PERCENT HYDROGEN IN ITS ANNEALING PROCESS.

WDRB Fox 41

In 1971—before Denny Crum, Jon Jory, the return of AAA baseball, and even the city's Riverfront Renaissance—the Louisville television landscape changed for good with the debut of WDRB Fox 41, the area's first new commercial station in more than a decade. ■ Originally operating out of a converted warehouse in Louisville's Butchertown neighborhood, WDRB enjoyed immediate success among viewers and advertisers alike by counterprogramming against its network-affiliated competitors with a daily diet of movies, children's shows, and syndicated comedies and dramas. And memorable on-air characters such as Presto the Magic Clown and the Fear Monger, *Fright Night*'s horror host, provided the station with an immediate identity that set it apart from its more traditional rivals.

The Evolution of a Leader

Before the end of the seventies, the original owners of WDRB sold the station to the Minneapolis Star and Tribune Company (later known as the Cowles Media Group), and Channel 41 took the first of many bold steps by moving its entire operation into brand-new quarters at Seventh Street and Muhammad Ali Boulevard in the heart of downtown Louisville in June 1980.

In the mid-1980s, WDRB changed hands again when it was acquired by Blade Communications, Inc. of Toledo (now known as Block Communications Inc.), one of the nation's most respected family-owned media companies, whose holdings today include three other television stations, numerous cable TV systems throughout the country, and the award-winning newspapers the *Toledo Blade* and the *Pittsburgh Post-Gazette*. With Blade Communications' acquisition, WDRB was positioned to become much more aggressive in enhancing its community profile, and immediately set out to do so through the production of more local programming. This commitment was perhaps best exemplified by the station's original agreement in 1985 to become the flagship station for all University of Louisville televised sports—an agreement that still exists to this day.

In 1986, new opportunities beckoned for WDRB when it became a charter affiliate of the then-new Fox Broadcasting Network. This

Clockwise from top:
The WDRB Fox 41 headquarters is located on Muhammad Ali Boulevard.

The *Fox News at 10* team includes (from left) Co-anchors Don Schroeder and Lauretta Harris, Sports Director Gary Montgomery, Meteorologist Tammy Garrison, and Business Reporter Bill Francis.

Fox in the Morning is co-anchored by Kevan Ramer and Elizabeth Woolsey.

marked the beginning of a broadcasting revolution that established the viability of a fourth over-the-air network, and ultimately produced such television classics as *The Simpsons, The X-Files, Ally McBeal,* and many more. Throughout the Fox network's existence, WDRB has consistently ranked as one of its strongest affiliates.

A Leader in News

Perhaps the most significant step in the evolution of WDRB came in March 1990, with the premiere of the station's first locally produced nightly newscast, *The News at 10*. At first seen only on weeknights, *The News at 10* was instantly successful, and soon expanded in two ways: from five to seven nights a week, and from 30 minutes to a full hour. Since undergoing a name change to *Fox News at 10*, the program has seen its audience grow, due to such factors as its earlier time period; its expanded, more in-depth format; and the inclusion of regular features, such as a nightly Business Report, that are unique among Louisville television news outlets. And the success of *Fox News at 10* hasn't been limited to good ratings; the newscast has been honored with many industry awards, including several Emmys, for excellence in broadcast journalism.

Following the success of *Fox News at 10*, the next logical step for WDRB was to expand its news offerings into other times during the day, and that step became a reality in October 1998 with the premiere of *Fox in the Morning*

and *Fox First News*. *Fox in the Morning*, which airs Monday through Friday from 6 to 9 a.m., is the area's only morning news show that offers 100 percent local content, while *Fox First News* follows the *Fox News at 10* formula for success by offering a midday newscast at 11:30 a.m., 30 minutes earlier than the competition.

WDRB expanded its broadcast activity in the Louisville market in 1994, when it agreed to operate a new station, WFTE-58, which is licensed to serve Salem, Indiana, and is affiliated with the UPN network. Since its sign-on, WFTE has attempted to follow WDRB's lead in providing unique community service, and that philosophy is perhaps best illustrated by the station's commitment to coverage of the southern Indiana sports scene—from Indiana Boys and Girls High School Championships to Indiana University football and basketball.

Giving Back to the Community

Together, WDRB and WFTE have devoted countless hours to such community groups and activities as Junior Achievement, the Center for Women and Families, the Louisville Zoo, the Kentucky Derby Festival, and the Kentucky Athletic Hall of Fame. And on-air public service campaigns such as Fox 411, Louisville's Most Wanted, Kentucky Crossroads, and Discover Southern Indiana have addressed local issues—and showcased local treasures—as only a fully committed member of the community can.

In 1971, WDRB started as a business. But now, in the 21st century, the station has evolved into much more: a vital part of a thriving metropolitan area, dedicated not only to achieving broadcast excellence, but also to enhancing the quality of life for everyone within its reach.

WDRB'S SKYFOX HELICOPTER AND SATELLITE UPLINK TRUCK PROVIDE VIEWERS WITH INSTANT ACCESS TO BREAKING NEWS.

CLEAR CHANNEL COMMUNICATIONS, INC.

With 832 radio and 19 television stations in the United States, equity interests in more than 240 radio stations internationally, and more than 550,000 outdoor displays in 32 countries worldwide, Clear Channel Communications, Inc. is a global leader in the out-of-home advertising industry and the largest radio group in the world. The San Antonio-based broadcasting company began its operations in 1972, and became a publicly traded company in 1984.

Serving the Louisville Community

Clear Channel Communications became a part of the Louisville community in 1986 when it purchased WHAS-AM, a news/talk radio station, and WAMZ-FM, a country radio station. Today, these two stations are the top-rated radio stations in the city. Since this initial acquisition, Clear Channel has added six other stations to its Louisville radio network, including WZTR-FM, WTFX-FM, WQMF-FM, WYBL-FM, WWKY-AM, and WKJK-AM. These stations were purchased in the mid-1990s, when the communications leader was just beginning to grow into a nationwide network. The Kentucky News Network, a statewide news service, is also a member of the Clear Channel family.

Currently, Clear Channel's radio network holds more than 60 percent of the Louisville listening market. "Clearly, our two leading stations, WHAS and WAMZ, have helped us establish that lead," says Vice President and General Manager Bill Gentry. "But we are also very strong in the rock market with WQMF, WTFX, and WZTR."

WHAS has served the Louisville community for nearly 80 years and ranks among the most-listened-to AM stations in America. WHAS has won numerous awards, including four Peabody Awards for Excellence, more than 100 Associated Press (AP) awards, and many others. The station serves as the community's number one news/talk outlet, and is one of the highest-powered stations in the nation. Powered by 50,000 watts, the station can be heard in 38 states around the country. WHAS Operations Manager Kelly Carls credits the entire crew with the station's success. WHAS features such legendary personalities as Milton Metz, who has been with the station for more than 50 years; Bob Sokoler on the morning team; and Terry Meiners in the afternoons.

WAMZ's Coyote Calhoun was the first on-the-air employee at the station, which today ranks as the city's number one radio station. As one of Louisville's most recognizable personalities, Calhoun has won many awards, including Country Music Association (CMA) Air Personality of the Year, Billboard Air Personality of the Year five times, Billboard Program Director of the Year four times, and many others.

Quality Programming

We keep listeners and advertisers with the quality of our programs, and by maintaining our local appeal and

WAMZ-FM became a Clear Channel Communications, Inc. station in 1986. Today, the company holds more than 60 percent of the Louisville market.

local values," says Gentry. "The radio personalities from all of our stations are the ones who make it happen."

All of the Clear Channel radio stations come together in one building, making it perhaps one of the most varied locations in the city. According to Metz, this diversity contributes to the company's success. "Walk down our corridor lined with glassed-in studios, and you'll see men and women delivering the news, talking sports, telling jokes, giving commercials, and chatting with dozens of callers on the phone—this demonstrates the diversity of programming involved," says Metz. "Seeing men and women in designer business suits, others in jeans or sweats, even shorts—they give you an impression of the range of personalities involved in our family of broadcasters."

Gentry adds that the stations' guest lists are also varied: "It's not unusual to see governors, head coaches, or U.S. senators featured in our studios. You never know who is going to be here. From the members of the wildest rock band to the governor, all kinds of guests are featured here in the studio."

ATTENTION TO LISTENERS AND ADVERTISERS

Clear Channel places a priority on listeners. For the variety of musical preferences in the listening population, the company develops creative broadcasting solutions. "We cover nearly every radio format, from news/talk to country, a variety of rock stations, and easy listening," says Gentry.

Clear Channel relies on extensive market research surveys to get a feel for listener preferences. And while the company is interested in supporting new formats, no new radio formats are put into the market without extensive testing. Listeners also benefit from the size of the Clear Channel group. Because Clear Channel can offer blocks of advertising for a particular region or for several stations at once, smaller stations that fill an audience niche can remain on the air, supported by advertising revenue from the entire network. The unique diversity of the audiences allows advertisers a deeper penetration into the market.

Clear Channel uses the Internet to expand its services as well. It offers detailed sports information and programs, and simultaneously broadcasts some programs via the Internet. Listeners can use E-mail to communicate with their favorite personalities and voice their opinions.

LISTENING TO THE COMMUNITY

Clear Channel recognizes the direct impact it has on the community it serves. The company has used its broadcasting to help many great causes in the community, including serving as a sponsor for Crusade for Children and St. Jude fund-raising events.

"More than just sponsoring any individual event, we have the pulse of the community, so we are responsive and proactive," says Gentry. "We try to be there first. Where there is a need is where we will focus our attention." From providing the best-quality programming to helping advertisers sell their products and helping the community it services, Clear Channel Communications will continue to be an important part of Louisville for many years to come.

CLEAR CHANNEL COMMUNICATIONS EARNS AND KEEPS LISTENERS AND ADVERTISERS THROUGH QUALITY PROGRAMMING AND A WEALTH OF RADIO PERSONALITIES.

Gordon Insurance Group

When Bernie Gordon founded Gordon Insurance Group in 1975, he was convinced that strategic thinking was the key to providing the best possible business insurance. That philosophy has guided the growth of the company, which has become one of the largest insurance agencies in the Louisville metropolitan area. ■ Effective insurance planning, like chess, requires the ability to see beyond the immediate threat and make the moves that will protect a client's family, business, and home against catastrophic loss. For more than 25 years, Gordon Insurance Group has worked as a grand master of insurance, providing its local, regional, and national clients with unparalleled service.

The Gordon Insurance Group provides a wide range of insurance and risk management services to businesses and individuals. In addition to property and casualty protection, the company offers life, health, and employee benefit programs, as well as self-insurance program administration.

Gordon Insurance Group represents only the top insurance companies and emphasizes maintaining effective relationships with a limited number of these companies. In doing so, the group has been able to establish associations exclusively with insurance carriers who have demonstrated the range of coverage the agency's clients require with a long-term, stable record in the marketplace. Because the Gordon Insurance Group produces a sizable percentage of business to those companies, the group is better able to influence underwriting criteria, rate structure, and loss history interpretation.

The respect shown to the Gordon Insurance Group is also a direct result of the quality of business it brings to the insurance carriers. As an agency, Gordon Insurance Group has one of the lowest loss ratios to be found in the region. This credibility in the marketplace translates into real savings and greater security for the firm's clients.

BERNARD GORDON IS THE FOUNDER AND PRESIDENT OF GORDON INSURANCE GROUP.

Sincere Concern

The success and performance of an insurance agency often comes down to how it reacts when a loss occurs. The Gordon Insurance Group encourages its clients to contact the agency first, after necessary emergency personnel have been called. The company has a separate claims department that can give prudent advice and helps clients act decisively to minimize loss, both in a crisis and after the event has occurred.

Because the aftermath of loss can be as damaging as the loss itself, the company reacts quickly to help clients collect all pertinent information, such as police reports and insurance company forms. The information is immediately sent to the insurance carrier to expedite payment.

As an agency, Gordon Insurance Group has attained its level of success by steady growth, retention of valued clients and employees, and expansion of professional services. All of these accomplishments can be attributed to the firm's mission: to provide strategic thinking in managing risk. The company's

GORDON INSURANCE GROUP IS LOCATED IN THE DUPONT PROFESSIONAL TOWERS BUILDING IN DUPONT SQUARE.

expertise in risk management goes hand in hand with its sincere concern for the best solution for its clients, making Gordon Insurance Group a continued leader in business insurance.

STRATEGIC PLANNING

At Gordon Insurance Group, a methodical approach to planning begins with a thorough understanding of a client's business and the dangers it may face. Before offering an insurance product, the agency provides a risk management team to work with each client. The team analyzes the client's business and assesses which factors affect protection. The analysis also includes determining a client's loss history and determining what can be done to help minimize future risk.

For Gordon Insurance Group, true service lies in the details. The agency employs expert staff members who stay informed of the latest policy terms, conditions, and underwriting criteria. They conduct an annual review of each client's protection, and look for gaps and deficiencies in coverage.

THE BEST COVERAGE POSSIBLE

Many insurance agencies rely solely on the companies they represent for underwriting expertise, accepting their conclusions without an internal analysis. However, Gordon Insurance Group uses its own in-house underwriters to review the work insurance companies use in their decisions, ensuring the information is accurate and the coverage is the best available.

Gordon Insurance Group's thoroughness in preparing applications and providing information about each client's operations assists the insurance carriers in quoting and writing coverage accurately. This attention to detail has won the agency the respect of the insurance companies it deals with, providing negotiating leverage on the client's behalf.

GORDON INSURANCE GROUP EMPLOYEES WORK AS A TEAM TO ENSURE THEIR CLIENTS THE BEST POSSIBLE COVERAGE.

FOR MORE THAN 25 YEARS, GORDON INSURANCE GROUP HAS WORKED AS A GRAND MASTER OF INSURANCE, PROVIDING ITS LOCAL, REGIONAL, AND NATIONAL CLIENTS WITH UNPARALLELED SERVICE.

Samtec, Inc.

Although it might not be a household name, Samtec, Inc. is a name to rely on for consumers in a variety of high-tech businesses. Samtec manufactures and sells board-level and micro interconnects in the $24 billion electronic interconnect market. These connectors are used in a variety of industries, including telecommunications equipment, information appliances, computers and peripherals, industrial controls, automation devices, medical equipment, and automotive and aerospace/defense products.

Samtec was established in 1976 in New Albany on the belief that exceptional service would open doors in a market that was already crowded with more than 200 competitors. Samtec found its niche with a concept called Sudden Service.

Innovative Customer Service

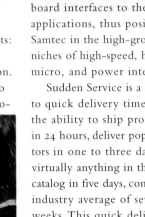

Sudden Service is comprised of three elements: Flex Design, Sudden Service, and Sudden Information. Flex Design allows customers to tailor a wide variety of board-to-board interfaces to their specific applications, thus positioning Samtec in the high-growth product niches of high-speed, high-density, micro, and power interfaces.

Sudden Service is a commitment to quick delivery times, including the ability to ship product samples in 24 hours, deliver popular connectors in one to three days, and ship virtually anything in the company's catalog in five days, compared to the industry average of seven to 10 weeks. This quick delivery time is especially important to engineers who are developing and testing new products.

Sudden Information—providing accurate design and order status information—is the final key element of Sudden Service. Samtec offers one of the most comprehensive Web sites in the industry, featuring personalized service and design centers for customers to retrieve data specific to their business and their company. Customers have access to their own pricing information, order status, product specifications, and lead times, among others.

Sudden Service is working. In 1976, Samtec grossed $170,000; in 2000, the company grossed more than $220 million and built a new corporate headquarters in New Albany that is three times larger than its previous building. This modern facility also houses Samtec's manufacturing, electroplating, inventory, shipping, and receiving facilities. Samtec employs some 900 associates in New Albany, and about 1,000 associates worldwide, with manufacturing facilities in Scotland, Singapore, and Malaysia, and sales offices in Germany, France, Italy, Sweden, Japan, Taiwan, Shanghai, and Hong Kong.

Winning Industry Acclaim

For associates at Samtec, customer service is a way of life. For five years in a row, Samtec has been ranked number one by Bishop and Associates, the leading connector industry consulting group. Bishop and Associates surveys companies—Samtec's customers—within the electronics industry to rank connector manufacturers on their service and performance. The survey ranks the attributes that are important to

FROM TOP:
SAMTEC, INC. WAS ESTABLISHED IN 1976 IN NEW ALBANY.

FOR ASSOCIATES AT SAMTEC, CUSTOMER SERVICE IS A WAY OF LIFE. FOR FIVE YEARS IN A ROW, SAMTEC HAS BEEN RANKED NUMBER ONE BY BISHOP AND ASSOCIATES, THE LEADING CONNECTOR INDUSTRY CONSULTING GROUP.

SAMTEC'S IN-HOUSE TESTING, DEVELOPMENT, AND PROTOTYPE CAPABILITIES REDUCE TIME TO MARKET.

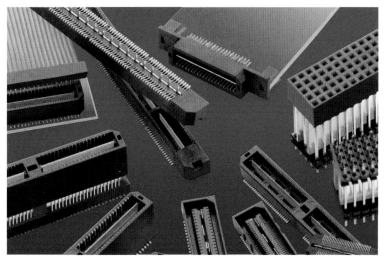

SAMTEC'S HIGH-SPEED, HIGH-DENSITY PRODUCTS ARE USED IN THE TELE-COMMUNICATIONS, COMPUTER, AND HANDHELD/INFORMATION APPLIANCE INDUSTRIES (LEFT).

SAMTEC'S VISION INTEGRATED ENHANCED WORKCENTERS (VIEW) IMPROVE QUALITY AND REDUCE MANUFACTURING COSTS.

customers when selecting connector suppliers, and compares the 34 top connector manufacturers. Samtec ranked number one in the five most important criteria in the survey: product quality, overall connector performance, meeting requested delivery dates, meeting acknowledged dates, and price competitiveness. The company also ranked number one in a dozen other survey categories.

This commitment to service has made Samtec the preferred supplier to companies such as Nortel, Rockwell, Motorola, Lucent, and IBM. "We're proud to win this award, especially for the fifth year in a row," says John Shine, president of Samtec. "In the connector industry, this is the equivalent of winning the J.D. Power quality award."

PRODUCT DEVELOPMENT

New product development is a major reason Samtec has grown an average of 25 percent each year from 1995 to 2000, compared to the industry average of 7 percent. Product development focuses on micro and high-speed interconnects. These are used in the growing telecommunication and data transmission markets. New products include connectors to maintain signal integrity in high-speed applications, and even smaller systems for use in pagers, cellular phones, and other handheld devices.

To ensure the development of new opportunities and new products, Samtec is constantly searching for and supporting innovative thinkers. For example, Samtec's Tiger Labs facility establishes research and development centers for entrepreneurs. The company also offers research and development partnerships, as well as venture capital opportunities.

DEDICATION TO ITS EMPLOYEES

Shine acknowledges that dedication to service, worldwide support, new products, and innovative information programs add to the company's growth. The real reason for the firm's success is Samtec's "dedicated associates who have put these systems in place and continue to manage and grow them," says Shine. "Daily, they make Sudden Service a reality for our customers."

Samtec's aggressive bonus program, liberal retirement plan, and Tiger Shares "phantom stock" options help keep associates focused on the Sudden Service goal. Samtec provides opportunities for compensation well above the market average, with a salary plus bonus based on company and personal performance. For example, when the company had exceptional sales in 1999, Samtec distributed a windfall bonus to all associates, in addition to the usual semiannual bonus. The company also supports an exceptional retirement plan, and makes 401(k) contributions three times better than the U.S. average.

Through an aggressive approach to better service, better products, and top-notch employees, Samtec, Inc. is prepared to lead the way in the interconnect industry well into the new millennium.

IN 1976, SAMTEC GROSSED $170,000; IN 2000, THE COMPANY GROSSED MORE THAN $220 MILLION AND BUILT A NEW CORPORATE HEADQUARTERS IN NEW ALBANY THAT IS THREE TIMES LARGER THAN ITS PREVIOUS BUILDING. THIS MODERN FACILITY ALSO HOUSES SAMTEC'S MANUFACTURING, ELECTROPLATING, INVENTORY, SHIPPING, AND RECEIVING FACILITIES.

Süd-Chemie Inc.

Süd-Chemie Inc. is a diversified specialty chemicals and industrial minerals company that employs more than 1,100 people, and has plants and mining operations throughout the United States. In March 2000, United Catalysts Inc. changed its name to Süd-Chemie Inc.—adopting the name of its parent company, Süd-Chemie A.G., of Munich—to send a clear message to the firm's customers that it is a single company able to provide unified services globally.

Süd-Chemie A.G. can trace its beginnings to 1857. The company originally began manufacturing a superphosphate fertilizer based on the pioneering work of scientist Justus von Liebig. The Süd-Chemie group, a recognized world leader in catalysts, adsorbents, and additives, operates more than 70 production and marketing companies throughout the world, providing jobs for some 5,200 people.

In 1974, Süd-Chemie A.G. acquired Girdler Chemical Inc. and, in 1976, Süd-Chemie A.G. acquired Catalysts and Chemicals Inc. (CCI). In 1977, Girdler and CCI merged to form United Catalysts Inc.

While Louisvillians may be familiar with the two plants located in Louisville, most people are not as familiar with the products that Süd-Chemie Inc. produces. Bert Knight, president of the Louisville facility, says, "Because we create very few products that an everyday consumer will purchase, it is difficult for people to easily relate our name to a specific product. But chances are, no matter where you live or what you do for a living, you use something made with or containing one of our products every day."

Süd-Chemie Inc. got its start making catalysts for the production of ammonia and hydrogen. Through acquisitions and new product creations, the company now operates four business units—SynGas, Refining and Petrochemicals, Specialty Chemicals, and Air and Gas Purification Catalysts. These names are not readily associated with any specific product; however, each business unit produces catalysts that touch consumers in their daily lives.

Chemicals That Help Create Products

The Catalysts Division manufactures catalysts in a variety of sizes and shapes, including pellets, tablets, spheres, rings, and unique geometric designs for specific applications. Catalysts can be made from a wide variety of chemicals, but they generally come in a solid form. The catalysts take on the job of increasing the rate and efficiency of a chemical reaction.

While average consumers may not be able to give a good definition of a catalyst, if they have sipped anything from a Styrofoam cup, they have experienced how a catalyst helps to create products. "A catalyst is any substance that speeds up or slows down a chemical reaction without being consumed during the chemical reaction," according to Knight. "Styrofoam is made of styrene and without the addition of a catalyst in the chemical reaction of those substances, there would be no styrene and no Styrofoam."

There are many different applications for Süd-Chemie's diverse products, and research and development of existing and new catalysts is an important function for the two manufacturing plants and laboratories located in Louisville. The catalysts produced in Louisville help make many everyday products such as Styrofoam, plastics, ammonia, and even steel. Ammonia is one of the world's highest-volume commodity chemicals. Approximately 80 percent of the world's ammonia output is used in fertilizer and explosives, and the petroleum industry uses industrial hydrogen to refine crude oil into fuels. Some other catalyst uses include specialized environmental products such as catalytic air filters and refined gasoline, and many of Süd-Chemie's products are produced for specialists in the chemical industry. In

Süd-Chemie Inc.'s corporate headquarters is located in Louisville.

addition, the company creates unique catalysts for many specialty uses.

Improving the Flow of Things

When a consumer puts on body lotion, or even paints a room, rheological additives help make the product he or she is using flow smoothly and look better. Süd-Chemie Inc.'s advanced processing plant in Louisville is the newest and most modern rheological product plant in the world. The plant produces numerous powders, gels, and polymers for use in water- and solvent-based products, including paints, coatings, cosmetics, inks, and sealants.

Süd-Chemie Inc. provides superlative technical support to aid customers in finding formulation solutions to their rheological challenges. These can include manufacturing ease, package stability, product appearance, and application characteristics. Thus, from the production plant through to product use, Süd-Chemie's rheological additives are key ingredients of quality consumer products.

Süd-Chemie's newest research and development catalyst testing laboratory was built in 1998.

Drying It All Out

People often wonder about the exact purpose behind all of those little packets labeled Do Not Eat that seem to fall out of new packages of everything from medications to shoes. Each package actually contains desiccants—a special clay or other agent such as activated carbon or silica gel—designed to pull moisture out of the air. "Without desiccants, billions of dollars a year would be lost to products ruined by moisture," says Knight.

While these packets in consumer products are the most common usage of desiccants, there are also many other uses. Canned goods would be completely ruined by rusted cans if they were not packed with desiccants during shipping. Desiccants also keep electronic equipment dry and safe during shipping. Other important uses include keeping military equipment and armaments safe during shipping and storage.

Süd-Chemie Inc. produces the highest-quality clay desiccant on the market today. The company's trade name, Desi Pak, has become synonymous with superior quality and performance. The special montmorillonite clay used in this product is mined in Arizona, and is processed and bagged in a state-of-the-art facility in Belen, New Mexico.

In addition to their use in desiccants, Süd-Chemie Inc.'s clays and minerals are major components in the fiberglass and tile used to make bathtubs, toilets, sinks, and even floors. The clay that supplies all of these needs is kaolin, which is mined in Georgia by a Süd-Chemie Inc.-affiliated mining plant, Albion Kaolin. These clays and minerals are so omnipresent that they appear in a huge variety of products, including tennis shoes, plastic containers, and tires.

Stopping the Itch

For years, many people have avoided the woods because of allergies to poison ivy, poison oak, or poison sumac. Now, thanks to a new product by Süd-Chemie Inc. called IvyBlock, they can once again enjoy a relaxing stroll through the woods.

IvyBlock is the first product to receive FDA approval for use in preventing poison ivy, poison oak, and poison sumac when applied before exposure.

Two new products called IvySoothe and IvyCleanse are also now on the market. While IvyBlock must be applied before contact, these two products are for use after contact with poison ivy, poison oak, and poison sumac. IvySoothe is a cream designed to relieve poison ivy rash and itch. IvyCleanse is a towelette to be used for removing poison ivy oil that can cause the rash and itching.

"This product line is an example of creative uses for our many different products. Originally our Clay and Minerals Division received a request from a researcher who wanted something to help workers avoid ivy-related rashes. We worked hard on a solution, and found one," says Knight.

A Safe Place

Süd-Chemie Inc. is committed to continuous improvement of the safety and health of its employees, the safety of those around the plant, and the environment. As a member of the Chemical Manufacturers Association (CMA), the company

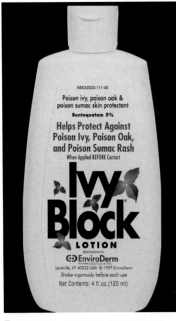

For years, many people have avoided the woods because of allergies to poison ivy, poison oak, or poison sumac. Now, thanks to a new product by Süd-Chemie Inc. called IvyBlock, they can once again enjoy a relaxing stroll through the woods (left).

The company's computer-controlled plants are staffed by highly skilled employees (right).

has adopted CMA's health, safety, and environmental performance improvement initiative, called Responsible Care. This initiative is made up of six codes of specific management practices to assure a safe work environment.

"Our people, plants, and products change. Therefore, we have to constantly update our employee training and our safety operations and equipment," says Knight. As part of providing a safe work environment, emergency response teams constantly prepare for any problems that might occur, including computer problems, fires, and gas leaks.

Partnerships with the Community

More than just trying to educate community members about their plant, Süd-Chemie Inc. employees have developed a partnership with the surrounding community. One example of the firm's local outreach is the Catalysts Kids Club, a program that was formed when a team of employees decided they wanted to take a group of 20 kids from a local adoption agency on a Christmas shopping spree. That onetime event has evolved into several yearly events, and has grown to include more than 40 deserving foster children and orphans.

Süd-Chemie Inc. has also developed a special partnership for the entire community around its plant. Again with a primary focus on children, plant workers also volunteer at the neighborhood school, Wheatley Elementary, and serve on its advisory panel. They also sponsor Kids Café, a program to provide hot meals for children in the area three nights a week.

Striving to be a good neighbor, Süd-Chemie Inc. works to keep the community involved through a Community Advisory Panel (CAP), made up of area leaders and concerned citizens. "We found that many people drove past our buildings, but didn't know what Süd-Chemie Inc. did. Our CAP program is a great way to get people from the community in to see what we do," says Knight.

In keeping with its own environmental awareness programs, Süd-Chemie Inc. also wants to provide a more attractive neighborhood. The company has achieved this goal by ensuring that all waste is out of sight and also by landscaping the area around the plant. In fact, the company's landscaping efforts won an award in 1999. But Süd-Chemie's cleanup efforts also go outside the walls of the plant. Employees have helped to organize, as well as volunteered for, major cleanup projects at local housing developments.

Always striving to be a good neighbor, Süd-Chemie Inc. works to create unique solutions for its neighbors just as it works to create unique solutions for its customers.

STRIVING TO BE A GOOD NEIGHBOR, SÜD-CHEMIE INC. WORKS TO KEEP THE COMMUNITY INVOLVED THROUGH A COMMUNITY ADVISORY PANEL (CAP), MADE UP OF AREA LEADERS AND CONCERNED CITIZENS.

WHILE LOUISVILLIANS MAY BE FAMILIAR WITH THE TWO PLANTS LOCATED IN LOUISVILLE, MOST PEOPLE ARE NOT AS FAMILIAR WITH THE PRODUCTS THAT SÜD-CHEMIE INC. PRODUCES. BERT KNIGHT, PRESIDENT OF THE LOUISVILLE FACILITY, SAYS, "CHANCES ARE, NO MATTER WHERE YOU LIVE OR WHAT YOU DO FOR A LIVING, YOU USE SOMETHING MADE WITH OR CONTAINING ONE OF OUR PRODUCTS EVERY DAY."

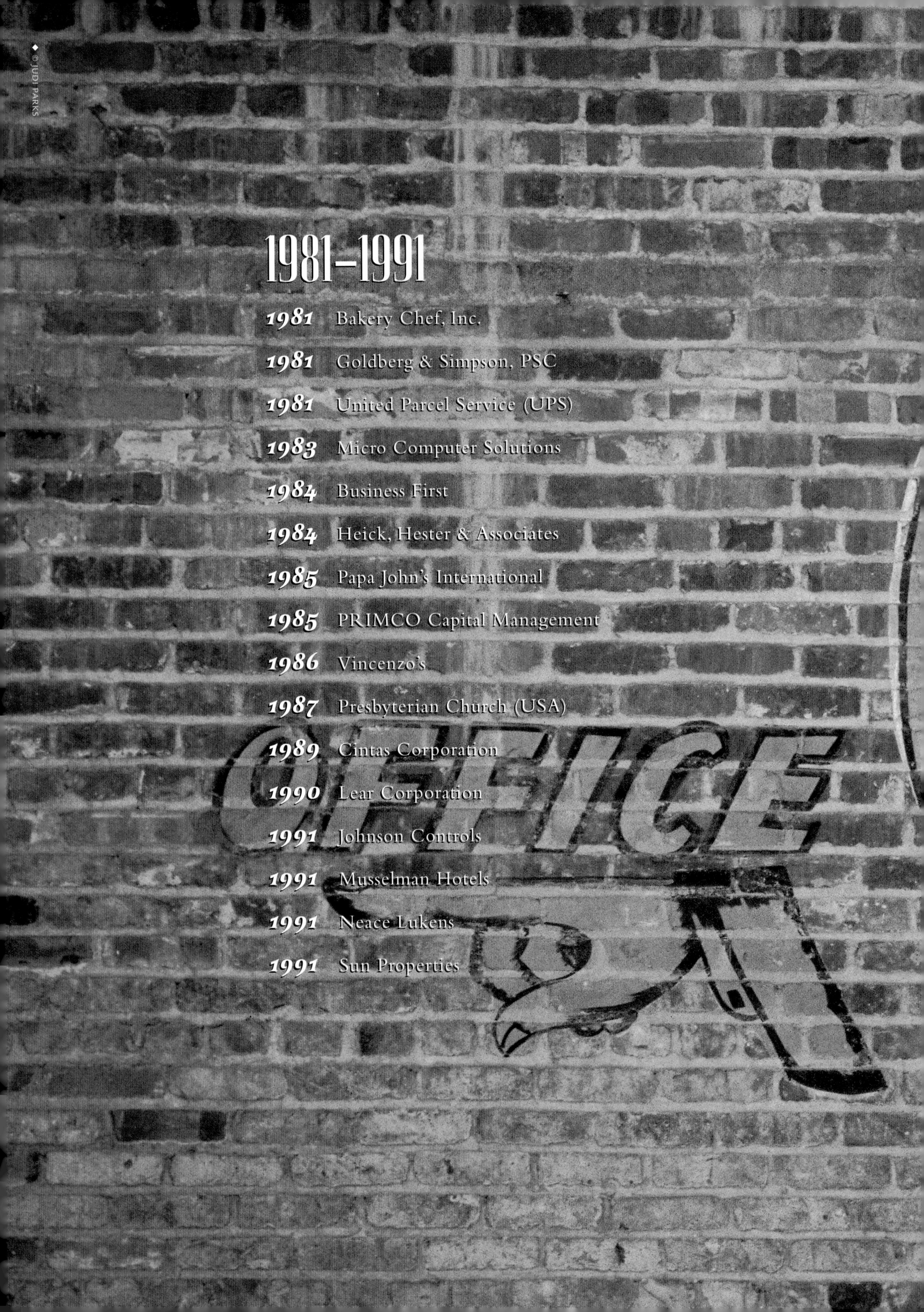

© JUDI PARKS

1981–1991

1981 Bakery Chef, Inc.

1981 Goldberg & Simpson, PSC

1981 United Parcel Service (UPS)

1983 Micro Computer Solutions

1984 Business First

1984 Heick, Hester & Associates

1985 Papa John's International

1985 PRIMCO Capital Management

1986 Vincenzo's

1987 Presbyterian Church (USA)

1989 Cintas Corporation

1990 Lear Corporation

1991 Johnson Controls

1991 Musselman Hotels

1991 Neace Lukens

1991 Sun Properties

Bakery Chef, Inc.

Louisville's Bakery Chef, Inc. is a leading supplier of biscuits, pancakes, waffles, French toast, muffins, dry mixes, specialty breads, and rolls. Although most consumers don't purchase the company's products at supermarkets, they can enjoy them at many food service locations, including some of the largest fast-food chains in the world. ■ Founded by Louisville entrepreneur Leo C. Culligan as Liqui-Dri Foods, Bakery Chef has been a part of the local economy since 1981. Originally, it was a small operation focused mainly on making dry foods—in particular, powdered soy milk blend sold to third-world countries. Culligan quickly realized that if he wanted to be successful, he would have to change his corporate strategy and expand his product line. So, in 1983, Liqui-Dri Foods began making dry biscuit mixes and marketing its products to fast-food chains across the country.

Liqui-Dri Foods' success attracted the attention of the Quaker Oats Company, which was attempting to get a foothold in the fast-food supply industry. The company purchased Liqui-Dri Foods in 1984. As part of the Quaker Oats family of companies, Liqui-Dri Foods expanded its product line and supplied dry mixes for pancakes as well as biscuits, and quickly became a leading supplier of dry mixes for the world's largest fast-food chains. The company now supplies the top customers in the food service, retail, industrial, and mass merchant segments of the food business.

BAKERY CHEF, INC.'S STATE-OF-THE-ART MANUFACTURING FACILITY IN LOUISVILLE, KENTUCKY, PRODUCES HIGH-QUALITY, VALUE-ADDED BAKERY PRODUCTS FOR THE FOOD SERVICE INDUSTRY.

Value-Added Bakery Products

During the late 1980s, the company followed its strategy by developing a line of value-added bakery products. The first products to appear under this banner were Ready-to-Bake (RTB) biscuits. "Our value-added bakery products are fully baked, frozen products that ensure quality, consistency, and ease of preparation," says Vice President of Supply Chain John Bischoff. "They help our valued customers save time, and more importantly, they assure them a high-quality and consistent product. Whether a customer is eating a biscuit at a particular fast-food restaurant in Los Angeles or one in Louisville, that biscuit will be of the same high quality."

Today's fast-food restaurants strive to provide faster service and better products, and Bakery Chef has stepped in to help them achieve that goal. "I believe there is a fundamental shift from people preparing from-scratch food to people preparing ready-to-bake food," says David Beré, chief executive officer. "As restaurants continue to serve bakery products, we can provide a higher-quality product at a lower price."

By 1990, the value-added bakery products were doing so well that the company built a $35 million facility in Louisville to increase production and meet the demand. And, in 1998, Bakery Chef added frozen pancakes to its value-added bakery line of products.

Reviving the Entrepreneurial Spirit

Bakery Chef moved back into its role as an independent business in December 1998, when the Quaker Oats Company decided to divest

DURING THE LATE 1980S, BAKERY CHEF DEVELOPED A LINE OF VALUE-ADDED BAKERY PRODUCTS. THE FIRST PRODUCTS TO APPEAR UNDER THIS BANNER WERE READY-TO-BAKE (RTB) BISCUITS.

BAKERY CHEF'S VALUE-ADDED BAKERY PRODUCTS, SUCH AS PANCAKES, ARE FULLY BAKED AND ENSURE QUALITY, CONSISTENCY, AND EASE OF PREPARATION.

the business and refocus on its core retail brands. Although some companies might have had a difficult time moving from a giant corporation back to a more entrepreneurial mode, Bakery Chef managed the transition smoothly.

Says Bischoff of the change: "We were really able to better focus on our top priority of exceeding the expectations of our key stakeholders: our customers, our employees, and our suppliers." In fact, Bakery Chef nearly doubled in size during its first year as an independent, privately held company.

Although Culligan retired during the mid-1990s, many of the key staff that helped Bakery Chef excel during its tenure as a Quaker Oats subsidiary remains with the company today. "We are very focused on our employees," says Bischoff. "In fact, the employees were the ones who helped rename the company. We held a contest for our employees to give our company a new name and had more than 200 suggestions."

Beré points out that the skilled Bakery Chef employees make the difference: "Our people are our chefs. The collective experience of our employees is what allows us to create excellent biscuits, pancakes, waffles, French toast, muffins, dry mixes, specialty breads, and rolls."

GIVING BACK TO LOUISVILLE

Bakery Chef continues its long-standing tradition of investing in the local Louisville community. It continues to expand locally, and recently invested more than $20 million to enhance its local manufacturing facilities.

Bakery Chef, along with its employees, also takes an active role in giving back to the local community through involvement with Metro United Way and the local chapter of the Ronald McDonald House, to name a few. "Our employees are always willing to help others less fortunate than themselves," says Bischoff. "Whether it is a walkathon for a national charity or collecting food for the local food bank, Bakery Chef employees are there to help."

FOCUS ON THE FUTURE

With emphasis on community involvement and superior-quality products, Bakery Chef is looking forward to continued growth in the future. "Acquisitions will be key for our growth," says Bischoff, "but our strategy will be to maintain our food service heritage. We want to continue to partner with our customers to assist in the building of their brands while establishing the Bakery Chef brand portfolio as well."

Bakery Chef has a long history in Louisville, and with its exciting growth strategies, there is no doubt the company will continue to be a dominant force in the value-added bakery products arena.

ON-LINE QUALITY ASSESSMENTS BY BAKERY CHEF TEAM MEMBERS ENSURE THAT THE MILLIONS OF PANCAKES PRODUCED DAILY MEET THE HIGHEST OF STANDARDS.

Goldberg & Simpson, PSC

Clients first in service and attitude: This has been the mantra of Goldberg & Simpson, PSC, a law firm located in Louisville, since the firm's inception in 1981, and continues to drive the firm earnestly into the next millennium. ■ According to Jonathan D. Goldberg, the firm's managing director, the primary principle upon which Goldberg & Simpson was established was to provide its clients with quality, competent representation at an affordable cost. "I think what makes Goldberg & Simpson unique is that we have experience in each of the legal disciplines that now make up the modern practice of law," says Goldberg. "We have interaction among almost every discipline of the law—from corporate law to labor law, from general litigation to bankruptcy law, from claims defense to domestic law. We have found that none of our varied practice areas ever works within a vacuum."

A Team Approach

The key element that makes this firm unique is the tremendous energy shared among all of our attorneys," says Steven A. Goodman, chairman of the firm's corporate practice group. "We can work together on any project—no matter how big or small—and we have a support system that you would typically see in firms much larger than ours."

In addition to being involved within the boundaries of the courtroom, the attorneys of Goldberg & Simpson utilize alternative forms of dispute resolution, such as mediation and arbitration. It has been the experience of several members of the firm that these modern alternatives to trial bring about a much quicker and more efficient conclusion to a variety of legal disputes.

"It is our goal to provide a quick, efficient, and fair resolution to all of our clients' disputes," says Director Charles H. Cassis. "We make every effort to get to know our clients on a personal basis, thereby achieving a comfort level with them that they do not receive anywhere else."

Corporate and business law have become vital areas for several of the firm's clients. Members of the firm's corporate practice group continually endeavor to provide positive and creative solutions to their clients' individual and business problems. Utilizing the group's team approach, there are very few legal matters that cannot be addressed in-house. Goldberg & Simpson can offer all types of services geared to the establishment, growth, financing, and succession planning of any size or type of business.

"We have the capability to handle the problems of any type of business, big or small," says Christopher M. George, another of the firm's directors. "We have the resources to deal with the largest corporations in the community, yet we can also provide the personal service and attention to assist any small-business owner."

Technological advances have permeated the business community as a whole over the past several

A team approach benefits Goldberg & Simpson, PSC's corporate clients by offering creative solutions to complex legal issues.

GOLDBERG & SIMPSON'S EXTENSIVE EXPERIENCE IN MANY LEGAL DISCIPLINES IS BENEFICIAL TO ITS CLIENTS.

years, and Goldberg & Simpson has embraced this technology to provide better services to its clients. The firm has developed and maintained its own computerized litigation support system, putting the firm on-line with the Internet and the technical revolution, and giving attorneys who are working with volumes of legal documents immediate access to a massive cataloging and retrieval system.

A COMMITMENT TO THE COMMUNITY

What makes Goldberg & Simpson special is its commitment to the Louisville community on both a civic and a charitable level. With so many of its attorneys being natives of the city, it is only natural that Goldberg & Simpson would build a record of achievement for supporting local charitable organizations. Extensive involvement with such diverse groups as the Fund for the Arts, Metro United Way, Leadership Louisville, Ronald McDonald House Charities of Kentuckiana, Catholic Archdiocese of Louisville, and Jewish Community Federation shows that Goldberg & Simpson sets the standards by which all law firms should be compared.

MITCHELL CHARNEY AND STEPHANIE MORGAN-WHITE, TWO OF THE FIRM'S DOMESTIC RELATIONS ATTORNEYS, REVIEW A POSITIVE OPINION FROM THE KENTUCKY COURT OF APPEALS.

"Each of our professionals has made a deep, personal commitment to the community at large," says Emily L. Lawrence, a director within the firm's estate planning practice group. The firm's attorneys are encouraged to select specific community organizations in which they have a particular interest and to devote their time and talents to these organizations accordingly.

This sentiment is echoed by Mitchell Charney, chairman of the firm's domestic relations practice group. "One of the pillars on which this firm has always stood is that as a member of the community at large, it is an absolute requirement that all our attorneys engage in several civic and charitable activities. It starts when people first come to the firm, and continues through their tenure at Goldberg & Simpson."

Clients first in service and attitude: To the attorneys of Goldberg & Simpson, this is far more than simply a motto—it is the primary manner in which they practice law.

United Parcel Service (UPS)

At the beginning of the city's history, the mighty Ohio River made Louisville a center for the transportation of goods. Many of the same people who traded goods on barges and boats throughout the area eventually settled here. However, the river became less important for commerce as other modes of transport developed. ■ Louisville today has the enviable position of once again being a center for the transportation of goods, but this time on a fast-paced, global scale. It is the sky above the city—not the river surrounding it—that has become Louisville's most important thoroughfare. This new avenue of global commerce is fueled by United Parcel Service's (UPS) jet aircraft traveling to and from the company's main air hub in Louisville.

When UPS began a small air operation in Kentucky in 1981 with only seven jet aircraft and a mere 300 employees, Louisvillians didn't know the important role UPS would play in their future. Today, UPS has become the largest private employer in the city and the Louisville air hub serves worldwide operations, sorting more than half a million packages each day that are then sent on to more than 200 countries around the world.

Trucks aren't the only vehicles in the United Parcel Service (UPS) delivery fleet.

Hub 2000

UPS' Louisville hub is the foundation for the company's global network. In 1998, the firm began an expansion of the air hub. The expansion, dubbed Hub 2000, will create up to 6,000 new jobs and house more than 2.7 million square feet in building space. It will be able to process up to 300,000 packages and documents per hour.

But Hub 2000 will be more than just an expansion. UPS has also included many technological advances, such as high-speed conveyors and smart labels read by overhead scanners that will speed up the processing of packages and documents. Automated systems are also in place to rapidly transmit customs information to expedite the movement of international shipments.

Because of Louisville's central location, more than 80 percent of the country can be reached in just a two- to three-hour flight. The technological advances of Hub 2000 will speed up the sorting process to ensure a speedy path for all UPS packages.

Hub 2000 not only hails a new era for UPS, but for Louisville as well. Many local experts predict that the major expansion will create an additional 8,000 jobs that will spin off from other companies expected to locate in Louisville because of UPS' presence. Many firms are already relocating to Louisville to take advantage of the transportation and distribution capabilities associated with UPS.

With all of its recent advancements, UPS is moving at the speed of business. While the Internet serves as the virtual connection to customers for many companies, UPS plans to be the physical connection that brings the goods to the customer's door. As E-commerce expands, UPS plans to be the premier carrier for these cutting-edge companies. And just as the Internet has connected the world, UPS plans to expand its international presence to become the global network for businesses around the world.

More Than Just An Air Hub

Although the air hub is the center of UPS' Louisville operations, the city is also

home to UPS Airlines headquarters, the company's international customer service center and flight training facility, and UPS Customhouse Brokerage. One of the most recent additions is UPS Worldwide Logistics' Technology and Logistics Center. This center serves as a central storage and shipping location for a wide variety of companies, including those in the telecommunications, biotechnology, apparel, and electronics industries. Customers are offered services ranging from simple inventory and warehousing services to technological services such as rapid repairs and technical diagnostics.

The technology center plays a major role in the area of personal computer repair. When someone's personal computer breaks down, that person typically contacts the computer manufacturer for assistance and may have to send the computer back to the manufacturer for repairs. Instead of being shipped to the manufacturer, the computer is actually sent to UPS Worldwide Logistics' center. There, highly trained technicians diagnose the problem and complete repairs, returning the computer to its owner in a minimal amount of time.

PARTNERSHIP WITH THE COMMUNITY

UPS has expanded well beyond its original role as a provider of basic package delivery. Today, the company offers a broad-based, customer-driven portfolio of transportation and logistics services—and Louisville is home to it all. But the bedrock of UPS' presence in Louisville is the partnership the company shares with city, county, and state government.

Probably the most unique aspect of this relationship is the Metropolitan College partnership that was created to supply UPS with the more than 1,000 part-time employees it would need to partially fill all the new jobs Hub 2000 created. Reflecting the business-friendly environment of this community, business leaders, community leaders, and top educational institutions came together to create a partnership called Metropolitan College. The University of Louisville, Jefferson Community College, and Jefferson Technical College offer courses designed to fit into the schedules of employees who work the midnight-to-4 a.m. shift at UPS. In addition to their salaries, UPS employees receive full college tuition to attend any one of the participating institutions.

From a small hub in 1981 to an integral part of the community today, the Louisville arm of UPS is continuing to grow into the new millennium. With a dedication to customer satisfaction and a commitment to the community, UPS will play a vital role in the Louisville economy for years to come.

UPS' DEDICATED EMPLOYEES ENSURE PACKAGES ARE DELIVERED ON TIME EVERY DAY (TOP).

HUNDREDS OF THOUSANDS OF PACKAGES MOVE THROUGH UPS' LOUISVILLE AIR HUB EVERY DAY (BOTTOM LEFT AND RIGHT).

Micro Computer Solutions

Keeping up with changes in information technology (IT) is one of the biggest challenges facing the business world. That's where Micro Computer Solutions (MCS) steps into the picture. Since its inception in 1983, MCS has provided the Louisville community with the services needed to support business information systems. ■ Tim Hollinden, founder and CEO of MCS, began his career in the field of computers while still a college student, working in retail stores selling personal computers. "It was the early 1980s and personal computers were really just popping up for the first time in Louisville," says Hollinden. "What I really saw, though, was the lack of service and training after the sale, and a lot of angry customers as a result."

But Hollinden also saw opportunity, and he decided to start his own business. Working out of his home, Hollinden embarked on a mission of providing complete computer service and training. The business grew so quickly that, by 1985, he had to move MCS out of his home to an office; training facilities were added in 1987. That rate of growth continues for MCS, which now includes five satellite offices operating in two states.

Single-Source Solution

Information technology has changed at a rapid pace since 1983, but MCS has managed to stay ahead. The first company in the Louisville area to offer a single-source solution, MCS provides training, technical services, software development, Internet services, and consulting for the local business community. Hollinden states, "No one competes with us in the breadth and depth of services that we provide."

MCS can provide businesses with a project manager to help design and maintain a complete information system—saving time, money, and headaches by eliminating a search for individual services. "Before a business buys or updates its computer system, they should consult with us," says Hollinden. "The experts at MCS can help businesses select the perfect computer system for them. Then, we train their employees and supply them with experts to handle their system's needs."

As a one-stop shop, MCS provides unbiased service regardless of vendor or platform. The company is the only cross-platform training center with programs for Microsoft, Novell, and UNIX. Linux training is also offered, and Macintosh computers can be serviced and integrated into new networks.

Services for Every Technical Need

MCS is perhaps best known throughout the region for offering training in application software, Internet and

Micro Computer Solutions (MCS) provides the best, most innovative solutions to meet the needs of its clients by delivering high-quality, up-to-date training (top).

MCS' new headquarters is located in eastern Jefferson County (bottom).

MCS HAS BEEN CHOSEN THE AREA'S TOP COMPUTER EDUCATION FACILITY, OFFERING HANDS-ON TRAINING FOR A VARIETY OF TECHNICAL AND APPLICATION CLASSES.

Web design, programming, and operating systems. Classes are designed for beginners and technical professionals alike. MCS is an Authorized Prometric Testing Center, addressing today's huge demand for IT-certified professionals. Training can be done at the MCS training centers, a client's business, or an off-site location.

The company also offers a wealth of additional services to supplement its training programs. The MCS staff is broken up into specific groups, each providing clients with custom solutions for their particular needs.

MCS' Technical Services Group provides a variety of services, from simple PC repair to network and data communications integration, and—when planning an entire network or additions to an existing network—always plans for the client's current use as well as for future growth. The consultants at MCS can implement E-mail and office automation as well as voice and video communication technologies. MCS is experienced in local and wide area networks, connecting multioffice locations. Internet and intranet connections can be fully integrated with help from the company's Internet Services Group.

MCS' Software Services Group can develop software specifically

for a particular client or project, or adapt current market software to fit a company's needs. The Internet brings its own requirements to this group. Today's applications must be Web enabled, and Web sites must be functional with any operating system.

Developing an Internet strategy is a given in today's marketplace. MCS has the ability, with its Software Services Group and Internet Service Group, to plan the next move a company should make regarding the future of e-commerce. Web site design and development are available through MCS' creative staff, which can integrate a company's Web site, databases, e-mail, e-commerce, and much more. MCS' Contract Group goes one step further, providing the human resources necessary to establish and maintain an effective information technology system. MCS can recruit, interview, hire, and train qualified people for a business. Staff can be placed for long or short terms, as each specific case may require.

MCS: THE PERFECT FIT

MCS understands the challenges that businesses will be facing in the new millennium. "The challenge is not simply to know what advanced technological tools are available," says Hollinden. "You need to understand how to adapt them to your business communication needs, and how to use them to your best advantage." MCS meets that challenge with its Fully Integrated Technologies (FIT). FIT involves a complex mix of platform configuration, network design, software development, technical services, Internet services, training, and consulting.

Although Hollinden is far from his days as a student, he still understands that providing needed technical services is a valuable commodity. For almost two decades, MCS has been the perfect fit between technology and business in the Louisville community.

MCS' TECHNICAL SERVICES GROUP OFFERS CERTIFIED TECHNICIANS WHO ASSIST WITH PROVIDING THE SOLUTION TO A CLIENT'S NETWORKING TECHNOLOGY NEEDS.

Business First

Business First HAS BEEN THE PRIMARY SOURCE OF LOCAL BUSINESS news in the Greater Louisville area for more than 15 years. Each week, the newspaper delivers in-depth, timely, balanced news and information to more than 60,000 readers in the seven counties that make up Metro Louisville. ■ Tom Monahan, *Business First* president and publisher, says the newspaper has a three-pronged mission: "first, to provide well-reported, breaking news stories about what's going on in the Greater Louisville business community; second, to provide information and other data to help area businesses be more successful; and third, to provide informed opinions and to give the business community a forum for discussing important community issues."

Coast-to-Coast Coverage

Business First began publishing on August 13, 1984. It was the fourth in the American City Business Journals' chain of local business newspapers. American City, based in Charlotte, North Carolina, operates 41 weekly business journals, the Street and Smith sports publications, and the Network of City Business Journals, a national advertising representation firm. American City Business Journals is a unit of Advance Publications, one of the largest private publishers in the world. Its Condé Nast division publishes such heavy-hitting magazines as *The New Yorker*, *Architectural Digest*, *Vogue*, *GQ*, *Parade*, *Vanity Fair*, and *Wired*.

"Being owned by American City offers the best of both worlds," says Monahan. "We're allowed to operate like a small business, making all the editorial and news decisions here in Louisville. Yet we have 40 'sister' papers with whom we can share information, ideas, and news tips."

The excellent coverage in *Business First* is supplemented by its Web site at www.bizjournals.com/louisville. Through this venue, readers can access some of the stories in the weekly paper, as well as brief daily news updates. Businesspeople also can link to any of the 40 other business journals in the American City network and get information and services they need to grow, manage, and operate their businesses.

Talent, Experience Key

Business First's success is due to its ability to attract and retain talented people. Founding publisher Mike Kallay, who had enjoyed a distinguished career at the *Louisville Times*, led *Business First* for 10 years before accepting another position in the American City chain. He was succeeded by Monahan, who has been with the paper since its start.

Monahan began his *Business First* career as a reporter in 1984, and served as editor and vice president for nine years before becoming president and publisher on January 1, 1995. More than half of *Business First*'s 33 employees have been with the paper for at least 10 years, including every member of the management team.

"We have been very fortunate to be able to have hired such talented professionals who chose to build their careers here," Monahan says. "Any business is only as good as its people, and we have very good people."

More Than the News

In addition to publishing breaking news stories every week, *Business First* offers its readers a variety of features, columns, data, and special publications. "Our goal is to provide the information that businesses need to succeed," says Editor Carol Brandon Timmons.

Among the most popular features in the paper are "Profile," "People on the Move," "Small Business Strategies," "Focus," and "Business Leads." "Profile" features a local business leader, and "People on the Move" documents new hires by area businesses. The "Small Business Strategies" section offers news and information of value to owners of small businesses. The "Focus" section targets specific industries, such as finance, health care, high tech, and

Business First HAS WON NUMEROUS JOURNALISM AWARDS FOR ITS COVERAGE OF THE LOUISVILLE-AREA BUSINESS COMMUNITY.

Business First PRESIDENT AND PUBLISHER TOM MONAHAN HAS BEEN WITH THE PAPER SINCE ITS FIRST ISSUE IN 1984.

construction. "Business Leads" is a compilation of statistical information gathered from area courthouses.

Business First's special publications include *Business Women First, Minority Enterprise, Internet Guide and Directory, Tech.First, At Home, Forty Under 40, Call to the Post,* and *The Book of Lists. The Book of Lists* is an American City signature piece that compiles a year's worth of the popular reference lists that run each week in the company's business journals.

MARKETING SPECIALISTS

Business First offers the perfect advertising medium for companies looking to connect with influential businesspeople who make buying decisions for their businesses and families. Unlike other media looking to just sell ads, Executive Vice President and Director of Advertising Maureen O'Meara says her sales associates develop long-term relationships with their customers and become part of that company's team.

"We work with clients to develop advertising campaigns that bring results," O'Meara says. "We're not successful unless our clients are successful."

GIVING BACK TO THE COMMUNITY

In addition to being a successful business, *Business First* has been a solid corporate citizen. Its employees serve on numerous civic and charitable boards, and donate thousands

of hours of their time. The paper has been honored for its support of various organizations, such as the Metro United Way, Business and Professional Women of River City, Small Business Administration, Junior Achievement, and Volunteers of America. "It would be hypocritical of us to write editorials urging businesses to give back to the community if we

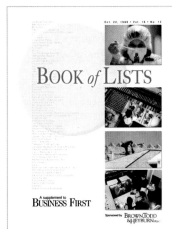

didn't practice what we preach," Monahan says.

For more than 15 years, *Business First* has been the voice of the Louisville-area business community and will continue to be for many years to come.

CLOCKWISE FROM TOP: *Business First* ADVERTISING EXECUTIVES WORK WITH CLIENTS TO DEVELOP MARKETING PLANS THAT WILL HELP THEIR BUSINESSES GROW.

THE KEY TO *Business First*'S SUCCESS IS A VETERAN STAFF OF DEDICATED PROFESSIONALS COMMITTED TO DELIVERING THE BEST IN LOCAL BUSINESS NEWS COVERAGE.

STAFF MEMBERS UTILIZE STATE-OF-THE-ART TECHNOLOGY TO PRODUCE A PLEASING DESIGN FOR THE PAPER AND ITS SPECIAL PUBLICATIONS.

THE ANNUAL *Book of Lists*, A COMPILATION OF THE WEEKLY RESEARCH LISTS THAT RUN THROUGHOUT THE YEAR IN *Business First*, IS THE PAPER'S MOST POPULAR SPECIAL PUBLICATION.

The Greatest City

Heick, Hester & Associates

Heick, Hester & Associates provides successful financial investment services to approximately 2,500 clients in the Greater Louisville metropolitan area. Founded in 1984, the firm manages more than $400 million in assets and is ranked as one of the top 10 financial advisers by American Express Financial Advisors, the nation's leading financial planning company. ∎

There are so many financial options to choose from today that without professional guidance, the choices may seem bewildering. New tax laws, volatile markets, concern for Social Security, rising health care costs, and longer life spans make financial planning critical to a secure future and peace of mind. "We have been successful because we have helped thousands of people and businesses make better financial decisions," says Carl Heick III, managing partner.

Business Partners for Success

Heick, Hester & Associates provides critical financial planning and investment advice for individuals and for employers and employees who rely on 401(k) plans. Most employees look to their employers for financial planning information and Heick, Hester has been at the forefront of helping employers fulfill these expectations. The firm has developed clear communications tools to help employees in 401(k) plans—managed by Heick, Hester for the employees' companies—to understand their many investment choices and the benefits of one versus another.

More than 50 employers already rely on Heick, Hester & Associates to provide efficient and successful management of their 401(k) plans. The companies, ranging in size from five to 2,000 employees, appreciate the services of Heick, Hester with minimal employer involvement. They've found that keeping employees informed about their 401(k) plans, as well as maintaining enrollment records and meeting the many state and federal reporting requirements, requires very little of the firms' time or effort.

"We work in partnership with our client companies' human resource management to offer the best range of options for their employees, as well as tailoring the education and information on their plan options for their best advantage," says Stephen Hester, managing partner. Heick, Hester & Associates has a team of eight financial planners dedicated solely to management of 401(k) investments and providing assistance to enrolled employees.

Individual, Family, and Estate Planning

For individuals and families, Heick, Hester & Associates provides comprehensive financial planning services, including education, retirement, and estate planning. "Today's economy is very different than it was just a generation ago," says Heick. "Needs have changed. Most people don't work for the same company all of their lives and, therefore, don't have the foundation of a long-term pension plan for their retirement.

Heick, Hester & Associates provides successful financial investment services to approximately 2,500 clients in the Greater Louisville metropolitan area. Founded in 1984, the firm manages more than $400 million in assets and is ranked as one of the top 10 financial advisers by American Express Financial Advisors, the nation's leading financial planning company.

Unfortunately, the changes in the Social Security structure have made it less of a sustaining income and more of a supplement to an individual's retirement portfolio."

Heick, Hester specializes in developing comprehensive financial plans geared to achieve each client's unique goals and objectives. Before developing a plan, each client's life goals and dreams and his or her current realities—income sources, mortgages, loans, and other debts—are thoroughly reviewed. Only after this detailed examination of the client's finances and objectives can a plan with properly diversified investment instruments be constructed to reach the client's goals.

"Many people do not understand their investment opportunities, and therefore don't make changes that could result in at least twice as much return on their investment," says Hester. The choices the typical consumer makes when selecting a retirement account are instructive: almost 50 percent of Americans keep their money in a fixed interest account with limited growth potential. "Fixed income accounts can have a place in diversified plans, but they severely limit growth," Hester explains. "We structure our clients' accounts to grow at much higher rates than the average American's rate of return."

Estate planning is a key part of Heick, Hester & Associates' work for many clients, ensuring that an individual's assets will be properly allocated after his or her death. "Our job in estate planning is to help coordinate the efforts of the entire team of professionals—the client's lawyer, accountant, and insurance agent—making sure the total estate plan will maximize the client's estate potential," says Heick. Without an estate plan, assets can be subjected to severe taxes that reduce the estate.

Heick, Hester planners focus on the client's financial situation and desires for distribution of his or her estate after death, working with the client's team of professionals to ensure trusts and charitable bequests are established and managed properly to achieve the client's intended purpose.

Winners of numerous awards and recognized for their expert management of client finances, both Heick and Hester rank in the top 10 of the 11,300 nationwide advisers associated with American Express Financial Advisors.

"The bottom line is helping our clients realize their financial goals," says Hester. "When our clients are successful, we are successful."

STEVE HESTER, CFP (LEFT), AND CARL HEICK, CFP, FOUNDED THE AGENCY TO HELP CLIENTS REACH RETIREMENT DREAMS BY BALANCING TODAY'S LIFESTYLE CHOICES WITH FUTURE SECURITY NEEDS THROUGH ONGOING FINANCIAL PLANNING WITH A TRUSTED, KNOWLEDGEABLE ADVISER.

PAPA JOHN'S INTERNATIONAL

John Schnatter has been in the pizza business since his teens, but in 1985 he delivered his last pizza for somebody else, picked up his business degree from Ball State University, and went home to Jeffersonville, Indiana. There, at age 22, he knocked down the broom closet in his father's tavern, installed an oven, and began delivering pizza out of the back of the bar.

From day one, Schnatter believed he could make a better pizza by using fresh dough and superior-quality ingredients.

Business at his father's tavern had been declining, and Schnatter was sure he could boost profits by selling pizza.

"Although the Papa John's concept started in 1985, the seeds of Papa John's culture were planted when I was just a youngster," says Schnatter. His father and grandfather had always drilled the importance of a strong work ethic into Schnatter.

From day one, Schnatter believed he could make a better pizza by using fresh dough and superior-quality ingredients. Schnatter's focus on quality has paid off. Today, with more than 2,400 restaurants throughout the United States and seven international markets, Papa John's is still focused on making a quality, traditional pizza. A limited menu of pizza, breadsticks, cheese sticks, and soft drinks keeps operations streamlined.

"You won't see chicken wings, salads, pasta, or subs on a Papa John's menu," Schnatter says. "We do pizza, and we always try to do it better than anyone else."

Quality Ingredients

Papa John's is a recognized quality leader among national pizza companies. Quality ingredients have such a high priority at Papa John's that Schnatter inspects them himself. Papa John's grows its tomatoes in a special California vineyard that produces an especially sweet tomato, which is perfect for pizza sauce.

Schnatter's zeal for quality ingredients is legendary. When he was having trouble finding quality olives, Schnatter went to the olive growers in Spain and asked them to find Papa John's a larger, better quality olive. He knew that a larger olive would make better slices for his pizzas.

To ensure that those quality ingredients make a superior-quality pizza, Papa John's has established a 10-point rating system to help the company's team members. That rating system identifies how the pizza should be cooked, how the crust and cheese should look, and even the proper placement of ingredients.

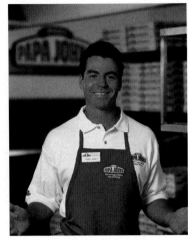

Although it is evident that Schnatter is the man behind Papa John's, he doesn't try to take all the credit. "Papa John's is not a one-man show. I owe my success to the people around me. The wisest thing I've ever done is surround myself with the smartest people in the industry," says Schnatter.

Core Values

Papa John's operates under six basic core values that are easily remembered as FASPAC—Focus, Accountability, Superiority, PAPA (People

At age 22, John Schnatter knocked down the broom closet in his father's tavern, installed an oven, and began delivering pizza out of the back of the bar.

Are Priority #1 Always), Attitude, and Constant Improvement. The corporate office and satellite restaurants work under these core values and are asked to make sure all actions fit with the core values.

"Our people are able to talk to any member of management if they feel something does not fit with our core values. I've visited restaurants before where a team member will tell me he doesn't think something quite fits with PAPA, and we will talk about changes that might need to be made," says Schnatter.

Always very hands-on, Schnatter has been known to pop into Papa John's restaurants throughout the country to ensure the pizzas are top quality. But he doesn't always remain a bystander. More than once, he has jumped into the kitchen to help make pizzas and deliver them when a restaurant has been shorthanded.

Papa John's efforts are paying off. Industry analysts around the world are awarding it recognition, including *Restaurants & Institutions*' choice as the highest quality national pizza chain for 1997, 1998, 1999, and 2000. And Papa John's was also ranked the number one pizza franchise by *Entrepreneur* magazine in 1999.

COMMUNITY LEADERSHIP

While Papa John's continues to improve its pizzas, the company strives to make a difference in local communities as well. It helped fund the Papa John's Cardinal Stadium at the University of Louisville, and also provided an endowment to the University of Kentucky for a basketball museum. Papa John's corporate presence has reached students throughout the nation with a scholarship program that provides more than $1 million a year for college scholarships.

Stepping up the company's level of community involvement, Papa John's has created a unique partnership with The V Foundation for Cancer Research, a charitable research and awareness organization named for legendary basketball coach Jim Valvano, who passed away from cancer in 1993. Papa John's helps support and advertise The V Foundation, whose funds will benefit selected research doctors as well as community cancer center programs.

Also, in an effort to help influence those on whom Papa John's most relies, the company began a partnership with the Future Farmers of America organization and is currently a major sponsor of its annual convention. "This is such a natural partnership for us. We need quality farmers to produce our quality ingredients," says Schnatter.

Many things have changed since John knocked down that broom closet, but one thing remains the same—Papa John's commitment to quality is still the driving force behind the company's success.

PAPA JOHN'S IS A RECOGNIZED QUALITY LEADER AMONG NATIONAL PIZZA COMPANIES.

PRIMCO Capital Management

AT THE HEART OF ANY SUCCESSFUL ORGANIZATION ARE HIGHLY talented and motivated people who have contributed to the firm's success. PRIMCO Capital Management is like many other firms, gaining its success from the quality of its people. ■ Founded by Vernon Hodge in April 1985 in Louisville, PRIMCO Capital Management has grown to become a global company and part of one of the largest investment management firms in the world. While PRIMCO may not be a household name, the company assures the financial security of thousands of households. Hodge retired in the early 1990s, but the foundation he laid continues to support the company.

Since its beginning, PRIMCO's primary focus has been on managing stable value funds that are designed to provide consistent, positive returns for clients. The people who utilize these stable value funds—through their employers' 401(k) plans or corporate savings plans—rely on PRIMCO to provide a secure income for their retirement or other needs.

These stable value funds are offered by many of the nation's largest employers, including Humana Inc., Ford Motor Company, Ashland, Inc., Sony Corporation of America, and Kellogg Company. PRIMCO's client accounts range in size from $27.9 million to $2.4 billion, and average approximately $297 million.

A Partnership for Success

In 1990, PRIMCO became a member of the INVESCO Group, a subsidiary of AMVESCAP, one of the world's largest independent investment management groups. As a member of the INVESCO Group, PRIMCO benefits from the full resources of being a part of AMVESCAP, which has more than $357 billion in assets, with 500 professionals in 25 different countries.

PRIMCO now has two locations that provide services for its clients worldwide. The main office is in Louisville, and the company's

GEORGE BAUMANN SERVES AS PRESIDENT OF PRIMCO CAPITAL MANAGEMENT.

PRIMCO EMPLOYS A DIVERSE AND TALENTED GROUP OF PROFESSIONALS DEDICATED TO PROVIDING FOR THE FINANCIAL FUTURE OF OTHERS.

other office is in Portland, Oregon. The combined staff of nearly 65 investment professionals specializes in all aspects of the fixed income and stable value marketplace.

LEADING THROUGH INNOVATION

In the 1980s, the stable value fund was the top investment choice of most savings plan participants. However, as the stock market soared, participants broadened their investment allocation to include more equity funds. "In 1985, nearly 70 percent of the market was invested in stable value funds; today, only about 18 percent of the market is in a stable value fund," says Randy Paas, vice president of PRIMCO. Amazingly, in this shrinking market, PRIMCO has managed to increase its assets from $9 billion in 1990 to more than $24 billion in 1999, a growth rate of nearly 300 percent.

As the market changed, PRIMCO became more innovative, offering different types of stable value investments to meet clients' needs. To adapt to the changing times of the early 1990s, PRIMCO created a new product. "Insurance companies had been offering guaranteed investment contracts (GIC), where a stated return is guaranteed by a financial institution such as an insurance company, but we saw a need for a new product that would better meet the needs of today's consumers," Paas says. "PRIMCO was an early pioneer in the creation of a security-backed GIC that has today become one of the leading products in the stable value fund market."

PRIMCO has continued to evolve and expand its investment universe to include separate account GICs, synthetic GICs, and actively managed fixed income. "Through our active management process, our clients have consistently realized higher yields on their fixed-income portfolios than they might have achieved on their own," says A. George Baumann, president and head of the fund management team.

"Our focus has always been on stable value fund clients," says Paas. "And our ability to meet our clients' needs through innovation is what has helped us grow."

SHARING GOOD FORTUNE

Sharing its good fortune has been a major part of the corporate culture at PRIMCO. "We truly believe in giving back to the community," says Baumann. PRIMCO has consistently been on the Metro United Way's Terrific 25 list, which recognizes smaller companies that make large contributions to the local charity drive. In 1999, PRIMCO and its employees donated more than $100,000 to the United Way, making it the second-largest contributor on the organization's Terrific 25 list.

The employees of PRIMCO believe in giving their time to the community as well. Each year, the company sponsors a Repair Affair house, where the employees supply the elbow grease and a few supplies to fix homes needing repair in low-income areas.

What began with one man as a local investment company has quickly grown into an industry leader in low-risk asset management, devoted to protecting not only the financial future of Louisville, but the futures of its clients around the country. Maintaining these goals, PRIMCO Capital Management will remain loyal to its founder's vision.

CLOCKWISE FROM TOP LEFT: ROWINA LYNCH IS A MEMBER OF PRIMCO'S CONTRACT ADMINISTRATION GROUP.

BEN ALLISON SERVES AS DIRECTOR OF MARKETING FOR PRIMCO.

KEVIN HORSELY, PRIMCO'S SENIOR CREDIT ANALYST, DETERMINES THE FINANCIAL STRENGTH OF INVESTMENTS.

PRIMCO'S TEAM OF PROFESSIONALS PROVIDE STABLE VALUE MANAGEMENT IN A CONTROLLED, DISCIPLINED MANNER. LING CHUI IS A MEMBER OF THE PORTFOLIO MANAGEMENT TEAM.

JAMES MYJAK IS A MEMBER OF PRIMCO'S INVESTMENT OPERATIONS GROUP.

Vincenzo's

Louisvillians don't have to travel to Italy to enjoy a traditional Italian meal. Instead, they can make the much shorter trip to Vincenzo's, an important part of Louisville's upscale dining scene since 1986, and enjoy an evening's meal with the Gabriele brothers. ■ Vincenzo and Agostino Gabriele were born in the Sicilian city of Palermo, and worked in the family's small restaurant, listening to tales of faraway countries from their father, a merchant marine. Both brothers were filled with desire to visit America in particular, which, according to their father, was a land of golden opportunity. While Vincenzo wanted to become a merchant marine like his father, Agostino found his true calling in the restaurant, where he was already creating wondrous delights even as a youth.

The Gabriele brothers finally came to America in 1969, not speaking a word of English and with nothing more than the clothes they wore. Their first destination was St. Louis, where an older Gabriele brother lived. Vincenzo worked as a dining room director in an award-winning St. Louis Italian restaurant and soon replaced his dreams of being a merchant marine with dreams of owning his own restaurant. Agostino, after a short stay with his brothers, left to travel Europe to hone his craft; he returned several years later to open Agostino's Little Place in 1976.

Vincenzo moved to Louisville in 1975 when a local restaurateur requested his services at an upscale Italian restaurant. In 1986, he opened his own restaurant, and within three months, he had convinced Agostino to come to Louisville to be his partner. "We really are a winning team," says Vincenzo Gabriele. "I love to work in the front of the house with our guests, and Agostino is truly a gifted chef."

A Traditional Italian Dining Experience

Vincenzo's offers guests a true Italian dining experience in an elegant setting. The typical meal at Vincenzo's might start with a traditional Italian risotto, or perhaps the chef's special, Crepes Agostino, a delicate blend of meats wrapped in a crepe and covered in a béchamel glazed marinara sauce.

The main course selections at Vincenzo's include a wide variety of some of the best pasta dishes in town. For heartier fare, veal, chicken, and fresh seafood are hand-selected and cut by Agostino, and prepared with a variety of tantalizing sauces. These delicious entrées can be accompanied by any of 300 different varieties of fine wine from Vincenzo's collection. Following the meal's main course, Vincenzo's guests will find dessert specialties such as a white chocolate mousse torte with a raspberry crème glaze or the classic Italian dessert, *tiramisu*.

JOHN NATION

Vincenzo Gabriele (right) moved to Louisville in 1975 when a local restaurateur requested his services at an upscale Italian restaurant. In 1986, he opened his own restaurant, Vincenzo's. Within three months, he had convinced his brother Agostino (left), a St. Louis restaurateur, to come to Louisville to be his partner.

Upscale Dining in Downtown Louisville

Located in the Humana building, Vincenzo's offers truly upscale dining in downtown Louisville. Vincenzo personally greets each guest and makes his way through the dining area to ensure that everyone is having a spectacular meal. Whether the guest is a celebrity passing through Louisville, a couple celebrating their anniversary, or even high school students on prom night, Vincenzo's ensures that each guest feels special.

The restaurant has served some famous people over the years, including Sylvester Stallone, Al Pacino, Diane Sawyer, Mikhail Baryshnikov, Hillary Clinton, Muhammad Ali, and Luciano Pavarotti, to name just a few. But Vincenzo's strives to make each guest feel just as famous as anyone pictured on the wall.

Vincenzo's has received numerous local, national, and international awards and ratings for its fine dining. The restaurant has earned several Best of Louisville awards and the Di Rona award for upscale, authentic Italian cuisine. However, one of the most prestigious awards for the Gabrieles came from their hometown of Palermo, where they were presented with a special award for advancing Italian culture in America through their restaurant.

Returning the Favor

The Gabriele brothers are not content to stand back on the sidelines and enjoy their good fortune alone; instead, they are more than willing to share with others. This gratitude extends to the restaurant's employees, many of whom have worked there since Vincenzo's opened. Such a low turnover rate is practically unheard of today, but the employees at Vincenzo's truly feel like part of a family. In addition, they participate in a profit sharing plan.

The Gabriele brothers' hospitality also reaches into the Louisville community. Each year, for example, the restaurant hosts a Thanksgiving dinner at the Cathedral of the Assumption for more than 300 of Louisville's homeless people. The brothers are actively involved in other ways to give back to the community, including the Home of the Innocents, the Lunch and Listen luncheons that benefit the Kentucky Opera, the Kentucky Baptist Home for Children, and the Girl Scouts, among other community service organizations. "We have been so fortunate," says Vincenzo. "We must return the favor."

What started as two brothers' dream for a better life in a new world has become a Louisville institution. An important part of Louisville history, Vincenzo's will continue to serve the local community for many years to come.

The main course selections at Vincenzo's include a wide variety of some of the best Italian dishes in town.

Presbyterian Church (USA)

The national offices of the Presbyterian Church (USA) serve as a resource for the denomination's more than 11,000 congregations across the country. The PC(USA) has had a strong history in Louisville and continues to impact the community both spiritually and economically. The church employs more than 700 people locally and directs the work of more than 800 staff members and volunteers involved in mission activities around the world.

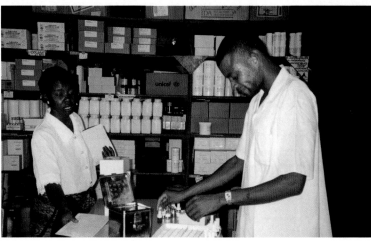

The pharmacy at Good Shepherd Hospital near Kananga, in the Democratic Republic of the Congo, is just one of the Presbyterian Church's many mission work efforts abroad. The 140-bed teaching hospital is supported largely by contributions from the Presbyterian Church (USA).

With roots that date back to the mid-16th century, the Presbyterian Church has contributed to the history and development of the United States since colonial times. The denomination split into two main groups around the time of the Civil War. In 1983, those two groups were reunited to form the Presbyterian Church (USA). When that happened, church leaders determined that they needed to find a new location for the church's national offices. More than 40 cities sought to become the denomination's new home.

In January 1987, a site selection committee recommended that the church offices be located in Kansas City, but Louisville successfully wooed church leaders away from that recommendation. In June 1987, more than 5,000 local residents turned out for a rally on Main Street to invite the PC(USA) to become part of the community. The successful effort was led by then-Mayor Jerry Abramson; John Mulder, president of Louisville Presbyterian Theological Seminary; and David A. Jones, chairman of Humana Inc. It was Jones who donated the two buildings, once owned by the Belknap Hardware Co., that became the church's headquarters.

Providing Services around the World

Today, Louisville is home to four of the six national offices of the PC(USA). The largest of these is the General Assembly Council, which coordinates a good deal of the church's mission work at home and abroad. Much of this work is done in partnership with native churches in more than 80 countries. The work includes humanitarian relief by the Presbyterian Disaster Assistance program, which responds to emergency needs with financial and volunteer support.

The Office of the General Assembly handles much of the denomination's administrative work and oversees meetings of the General Assembly, the church's highest governing body. The Presbyterian Publishing Corporation produces and markets books and other resources for congregations and the academic community. The Presbyterian Investment and Loan Program provides congregations with needed funds for building projects. A fifth national office—the Presbyterian Church (USA) Foundation, located in Jeffersonville, Indiana—manages financial investments and helps raise money to support the church's work. The sixth national office, the Board of Pensions, is located in Philadelphia.

The PC(USA) is a global enterprise, much of which originates with and is coordinated by those who work for the church in Louisville. With programs ranging from social justice to youth ministry, the PC(USA) strives to meet the spiritual and human needs of people throughout the United States and around the world. Its work in the areas of ministry, health care, education, and human development help the church fulfill its twin purposes of spreading the gospel and improving human conditions.

The Presbyterian Center on Witherspoon Street houses four of the denomination's six national offices.

SUN PROPERTIES

SUN PROPERTIES, ONE OF LOUISVILLE'S NEWEST COMMERCIAL development and property management companies, is built on more than 40 years of combined experience from its managing partners, Breck Jones and John Hoagland, who founded the company in 2000. "We saw the need for a new type of company—one that really puts the tenant's needs first," says Jones.

SUN PROPERTIES SHINE™

With each of its properties, Sun strives to live up to its mission of providing "superior service, unsurpassed cleanliness, and a nurturing environment for our clients." To that end, the firm has several programs in place to assure that its service really stands out from the rest. "We are really focusing on that extra level of service that tenants typically don't find with other commercial management companies," says Hoagland.

Unsurpassed cleanliness is a top priority at Sun Properties. If the exterior of a building needs painting or the interior needs daily cleanup, Sun Properties gets it done. Each of the firm's properties has a dedicated, on-site maintenance staff member (noticeable by his or her crisp, clean uniform) to handle all daily property needs.

Sun Properties also maintains standard response times to ensure quick repair for those problems that may occur with any building. "We pay special attention to the little details," says Jones.

NURTURING CLIENTS

"We want our clients to feel nurtured," says Hoagland. "We do more than just manage from a distance. Either Breck or I will visit each property daily."

Clients can choose from a variety of commercial properties, including A Commerce Center, which has 190,000 square feet on 17 acres; Professional Towers, with 157,000 square feet in the heart of the Dupont Circle Medical Center; and Embassy Square, with 125,000 square feet situated in an 8.5-acre, garden-style office park setting. Additionally, Sun Properties offers prospective clients build-to-suit opportunities in industrial and office properties.

By offering high-quality, personal service, Sun Properties ensures that each of its properties—old or new—has a special quality that makes it a comfortable place to work.

SOME OF SUN PROPERTIES' MOST NOTABLE PROJECTS INCLUDE (CLOCKWISE FROM TOP) DUPONT PROFESSIONAL TOWERS, EMBASSY SQUARE OFFICE PARK, AND A COMMERCE CENTER.

CINTAS CORPORATION

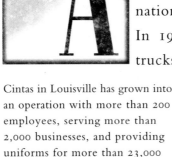

ALL TYPES OF BUSINESSES IN THE LOUISVILLE METRO AREA—FROM small service companies to major corporations—rent, lease, and purchase Cintas Corporation uniforms. Although Cintas is a national chain, Kentucky has been a vital part of its success. In 1989, Cintas established its service in Louisville, sending trucks down from Cincinnati. Starting with only two people, Cintas in Louisville has grown into an operation with more than 200 employees, serving more than 2,000 businesses, and providing uniforms for more than 23,000 people. The Louisville branch is one of the top 25 branches of the national chain and continues to grow at a phenomenal rate.

Statewide, there are more than 1,500 Cintas employees, with three service locations and four manufacturing facilities, making Cintas an integral part of Kentucky's economy. Nationally, Cintas ranks as a Fortune 1,000 company, and was listed as one of the 400 Best Performing Big Companies in the United States in the inaugural Forbes Platinum List. The company covers 190 of the top 200 U.S. markets, and estimates sales of nearly $2 billion in 2000.

EASE OF MIND

More and more employers are choosing uniforms as a key part of their corporate philosophy, especially today's service-oriented companies. "An employee who wears a clean, crisp, attractive uniform is always viewed as more professional and reliable than someone in ordinary clothes," says Greg Emrick, general manager.

Of course, today's uniforms are not the work suits of yesterday. Employers can choose attractive uniforms in a wide variety of styles, including casual polo shirts with company logos and coordinating slacks, shorts, and skirts. "Uniforms also help support a teamlike atmosphere, and build camaraderie and loyalty among coworkers," says Emrick. "And companies such as pest control, chemical, and research companies rely on our uniforms to protect their employees from exposure to soil, chemicals, or other contaminants."

The Cintas strategy is to provide ease of mind for its clients. It does that by managing the hassles associated with providing clean, well-fitted uniforms for employees. Cintas goes on-site to take measurements for the garments requested. The rental division supplies the employee his or her own specific set of clothing, which is bar coded for that particular person. A two-week supply of clothing is typically provided.

As a part of the rental service, the clothing is picked up and laundered at regularly scheduled intervals, repaired if necessary, and returned to the employee. The uniforms are continually updated for wear, tear, and size changes. Maternity uniforms are even provided for women who are expecting. "People who hire us can focus on their business, and let us take care of the uniforms," says Emrick. Ford, UPS, DuPont, Colgate-Pamolive, Beach Mold & Tool, and Publishers Printing are just a few of the local companies that are serviced by Cintas.

FOCUS ON CINTAS PARTNERS

The leadership at Cintas believes its corporate culture has been the driving force behind its success. And at

CLOCKWISE FROM TOP RIGHT: CINTAS CORPORATION PROVIDES HIGHLY SPECIALIZED SERVICES TO BUSINESSES OF ALL TYPES—FROM SMALL SERVICE COMPANIES TO MAJOR CORPORATIONS THAT EMPLOY THOUSANDS OF PEOPLE.

CINTAS' EXCEPTIONAL GROWTH MAKES THE DEVELOPMENT OF WELL-TRAINED PERSONNEL A HIGH PRIORITY.

CINTAS DESIGNS, MANUFACTURES, AND IMPLEMENTS CORPORATE IDENTITY UNIFORM PROGRAMS FOR CUSTOMERS THROUGHOUT NORTH AMERICA.

the very heart of the Cintas culture are its partners—the employees who day in and day out make it all happen. Nationally, the Cintas Corporation has grown for 30 consecutive years. In that 30-year period, sales have grown at a compound rate of 25 percent. A $1,000 investment in 1983, when Cintas went public, would be worth $47,000 today. "Our partners are the keys to that success," says Emrick.

Richard T. Farmer, founder and chairman of the board, sees employee involvement as an essential element of the company's continued success: "It is important to use the mind of every partner in the organization to recognize the potential of every human being." To this end, the managers are expected to listen to the partners who are doing the work on the front line. By telling the managers how to make the process run more smoothly, efficiently, and profitably, the partners can enable them to stay involved in the everyday operation of the plant.

Training is a vital part of Cintas' culture. The Service Team provides weekly training on customer service issues, some as simple as reminding employees to smile. Another example of Cintas' focus on well-trained, happy, and knowledgeable employees is the Cintas College. The leaders of the Louisville division gather monthly with groups of partners and talk with them about the culture and morale of the company. Through these meetings, the leadership enforces the idea that the employee's job is worthwhile and a big contributor to the success of the company.

Cintas realizes that another key to success is high employee morale. Cookouts and social events are often held for all 200 employees, while the daily work environment is friendly and fun for everyone— from management to workers on the line.

In addition to giving back to its employees, the company makes every effort to give back to the Louisville community, supporting University of Louisville athletics, United Way, and Fund for the Arts. But perhaps one of the firm's most important contributions is what it does best—providing clothes. Flood and hurricane victims around the world have benefited from donated clothing from Cintas.

From its growing presence in Kentucky to its motivating work environment and community involvement, Cintas Corporation brings a lot more than just clean uniforms to Louisville. The company's continued dedication to its employees and customers alike will assure its success for decades to come.

CINTAS PROVIDES RENTAL SERVICES FOR UNIFORMS, ENTRANCE MATS, AND FIRST AID AND SAFETY PRODUCTS FOR THOUSANDS OF COMPANIES THROUGHOUT KENTUCKIANA.

"EXCEEDING OUR CUSTOMERS' EXPECTATIONS" IS THE GOAL OF EVERY CINTAS EMPLOYEE.

Lear Corporation

The Lear Corporation began in 1917 as a small operation that supplied seat frames for the automotive industry in Detroit. At the time of its founding, Lear was known as American Metal Products and was one of the first companies to provide products for the automotive industry. Initially, the company provided only seat frames, but when Ford Motor Company introduced the Mustang in 1964, Lear manufactured complete seats for the first time. For several years, the Ford Mustang was the only car to boast complete seats by Lear.

In 1966, American Metal merged with Lear Siegler and began to grow, slowly expanding its presence in the automotive market. Since its beginnings as American Metal, the company had remained focused on automotive seats; however, during the 1990s, Lear expanded into complete automotive interiors, and the company saw dramatic growth in this market.

Today, the Lear Corporation is one of the world's leading full-service automotive suppliers, with complete interior design, engineering, testing, and manufacturing capabilities. The company's complete interiors include seats, door panels, instrument panels, headliners, and carpet. Worldwide, the Lear Corporation, headquartered in Southfield, has more than 100,000 employees in more than 300 facilities in 33 different countries.

Manufacturing and Safety Testing

Louisville became home to one of the Lear Corporation's most productive plants in 1990. The Louisville plant makes seats and other interior components under the just-in-time manufacturing process for Ford.

Through a special broadcast system that connects both facilities, the Louisville Ford plant orders seats and interior components from the Lear plant as the automobiles come down the line. By the time the Ford workers have completed the body of the automobile, the Lear seats and interiors have arrived; the seats are then placed into the vehicles without a disruption in

The Lear Corporation is one of the world's leading full-service automotive suppliers with complete interior design, engineering, testing, and manufacturing capabilities.

the assembly line manufacturing process at Ford.

In addition to its manufacturing facility, the Louisville plant also houses a laboratory to test the safety and comfort of Lear's seats and passenger restraints. Although safety is a primary part of the group's research, comfort is also an important factor. "Because the characteristics of foam and other components can vary, we have a testing facility on-site to test seats to ensure maximum safety and comfort for the occupant," says Wayne Allen, manager of the Louisville plant.

Quality Automotive Supplier

Lear has a well-earned reputation for providing some of the highest-quality products in the industry. In fact, its reputation has earned the company the premier job of manufacturing seats for the pace car of the Indy 500. Although Lear employees are especially proud of this particular project, they strive to ensure that all Lear customers receive the same quality and attention.

As quality regulations in the industry continue to increase, the Lear Corporation continues to achieve the highest standards ratings. While ISO 9000 certification is common to many similar facilities, the Louisville facility received the QS 9000 rating, a rigorous quality standard that required a two-year certification period set up by a third-party administrator. In addition, Lear was honored with a Commonwealth of Kentucky Quality Award for its quality manufacturing process.

"Our strength is our employees," Allen points out. "Without the commitment of employees, we could not have achieved the QS 9000 certification and would not have been recognized by the Commonwealth of Kentucky."

Bringing Louisville More

We want to do more than just build seats," says Allen. "We want to give back to the community." Lear strives to accomplish this goal by involving its employees in many different types of community giving. The company has made substantial donations to United Way, and over the last several years, the organization has awarded Lear for its significant increases in giving.

Another approach to community giving is the Adopt-A-Child program that sponsors more than 40 kids and their families at Christmas each year. Lear provides a holiday luncheon for the Adopt-A-Child families, who dine together with the Lear employees. The 40 children then receive a visit from Santa Claus, who provides gifts of toys and clothes to each child. "Our employees work hard to make this all possible," says Allen, "and we all enjoy being able to make a child's Christmas merrier."

Growth for the New Millennium

When Lear came to Louisville in 1990, the company opened its plant with only 50 employees. By 1999, the plant had more than 500 employees and had outgrown the old facilities. By mid-2000, Lear had moved into its new manufacturing plant, a 150,000-square-foot, state-of-the-art facility in the Eastpoint Business Center. The new plant allows for even more growth and could employ more than 600 by the end of 2001.

Lear's continued growth at this facility shows the company's commitment to both improving its market share, as well its commitment to the Louisville community. With a history of quality and the dedication of its employees, Lear will continue to be an industry leader for decades to come.

Lear's Louisville plant makes seats and other interior components under the just-in-time manufacturing process for Ford.

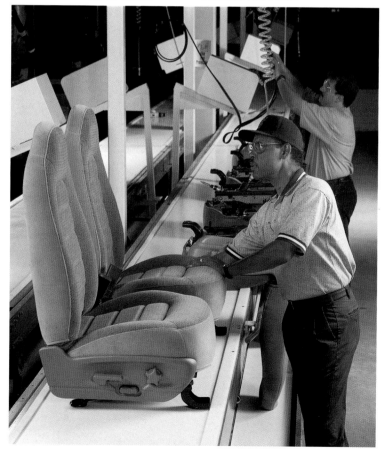

Johnson Controls

During his tenure as a professor at the State Normal School in Whitewater, Wisconsin, Warren S. Johnson received a patent for the first electric room thermostat in 1883. His invention launched the building climate control industry and was the impetus for a new company. By 1885, his entrepreneurial enterprise—known as Johnson Electric Service Company—was manufacturing, installing, and servicing automatic temperature regulation systems.

Throughout the years, the company has made many improvements in the field of automatic temperature control, such as constructing the industry's first minicomputer for building control in 1972. The company changed its name to Johnson Controls in 1974, and began a series of acquisitions and mergers aimed at expanding its services.

Johnson Controls initially entered the automotive industry by acquiring a company that manufactured car batteries. However, its most fortuitous move into the business came in 1985 when it purchased Hoover Universal, Inc., a manufacturer of automotive seating. Today, Johnson Controls has become the world's largest manufacturer of complete automobile seats, with manufacturing plants on five continents, and has secured a leading role in the automotive seat manufacturing business. Its worldwide locations can also supply the complete automotive interior for a variety of vehicles, including headliners, consoles, door panels, and instrument panels. In 1991, Johnson Controls opened a plant in Shelbyville, Kentucky, to supply Ford's Kentucky Truck Plant with seats for the vehicles it produces.

Just-in-Time Service

Supplying seats for vehicles produced at the Kentucky Truck Plant, the largest Ford plant in the world, requires detailed planning. Each seat manufactured by the 450 employees at Johnson Controls' plant is specially ordered for each Ford truck as it is being sent through the production line. Guaranteed to provide just-in-time service, Johnson Controls receives the order for each seat just hours before the seat is due at the Ford Truck Plant.

These tasks are accomplished through the help of an intricate broadcasting system. A Ford employee sends over an order for the specific seat needed for each truck. Johnson Controls then makes the seats and loads them into a truck to be shipped to Ford within a specific time frame—usually three and a half hours. When the employees at the Ford Truck Plant unload the shipment, the seats are all in the order of the trucks coming off the plant's assembly line. Not only does the seamless flow of production ensure a quality product that easily moves into the Ford assembly process, but it eliminates the need for Ford to keep a supply of seats on hand, which can save the company thousands of dollars a year in storage costs.

The Importance of Safety

Over the last decade, Johnson Controls has been dedicated to developing comprehensive research, development, design, engineering, and testing capabilities. This broad expertise is giving automakers and consumers seat systems with improved comfort, safety, and technology.

In addition to ensuring that its seats perform well in crash testing, the company also researches innovative technologies that increase safety. One such example is the pretensioner seat belt, which provides full restraint while being integrated into the actual seat and not the vehicle.

Quality Employees, Good Neighbors

Worldwide, Johnson Controls recognizes the need for businesses to be responsible community members,

Johnson Controls opened its Shelbyville, Kentucky, plant in 1991.

ASSOCIATES ASSEMBLE BENCH SEATS ON ONE OF EIGHT PRODUCTION LINES.

and the Shelbyville plant—the second-largest employer in the city—is a shining example. "We have a great relationship with the community, and we also have a great relationship with our employees," says Operations Manager Brenda Leggett. "Our employees are well rewarded and enjoy the work environment."

Johnson Controls provides both financial support and employee volunteers for a variety of local community service organizations. A team of management- and non-management-level employees completely organized the Kids ID campaign, a program that provided photo and fingerprint identification cards of children for their parents. Employees at Johnson Controls took photos and fingerprints at the local Wal-Mart and in several local banks.

There are many other community programs that receive the company's assistance. For example, employees often volunteer in the Shelby County school system's after-school reading program. Johnson Controls also offers job skills training programs for teachers from Shelby County schools, which they, in turn, teach to their students.

During the holiday season, employees regularly sponsor needy families, and provide gifts and meals. The firm's employees also contribute to local March of Dimes and United Way efforts, and are an ever present sponsor for local youth and adult sports teams.

"It is the quality employees we have that allow us to not only do quality work, but to be major community sponsors as well," says Leggett. "Throughout the Johnson Controls worldwide network, we are known as one of the best plants. And we continue to win awards that prove it."

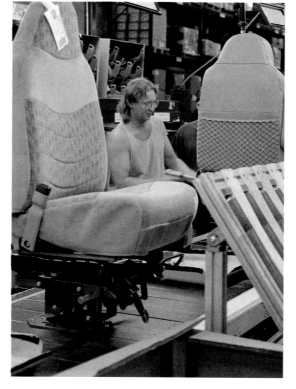

JOHNSON CONTROLS IS THE WORLD'S LARGEST PRODUCER OF AUTOMOBILE SEATS. ITS SHELBYVILLE, KENTUCKY, FACILITY SUPPLIES FORD'S KENTUCKY TRUCK PLANT.

Musselman Hotels

When Tom and Chester Musselman went into the hotel business in 1991, it was a very natural transition for the Louisville natives. Although they had never managed a hotel, their family had owned the land on Fern Valley Road where the Thrifty Dutchman hotel had been located for more than 20 years. ■ When the Thrifty Dutchman's managing company decided to sell the hotel in 1990, Tom Musselman Sr.—Tom and Chester's father—bought it and took over management duties himself. What began as an idea to own one hotel soon became a Musselman family business. Tom and Chester Musselman saw this as a great business opportunity and changed careers to become hoteliers.

In 1991, the Musselmans built their first new hotel, a Days Inn located in Ashland, Kentucky. That hotel is still one of the city's prime business hotels, offering many conveniences for business travelers.

Moving into the 21st century, Chester, president, and Tom, vice president of development, have carried on their father's tradition of excellence in service and management.

Built on Outstanding Service

Currently, Musselman Hotels owns and manages 22 hotels in Kentucky, Indiana, and Tennessee. The company's properties include a wide number of nationally recognized chains, including Marriott, Holiday Inn, and Comfort Inn. With the exception of the original Thrifty Dutchman in Louisville, Musselman Hotels has researched, built, and managed all its locations.

"We have been so successful because of our customer-focused philosophies," says Melanie Kelley, director of sales and marketing. "Choosing a good location and a good franchise are both important ingredients to success, but we believe our outstanding managers and employees are what really make our hotels different."

Musselman Hotels is focused on providing quality service in many ways. Each of its hotels has a distinct focus, and the corporate office works with hotel managers to determine each hotel's specific needs. "We have focused primarily on the business traveler," says Kelley. "Most of our facilities have meeting rooms and incorporate extra features like business centers and exercise rooms that are great for

The Second and Main Lounge, located in the lobby of the Courtyard by Marriott, has a panoramic view of the hustle and bustle of downtown Louisville.

THE COURTYARD BY MARRIOTT IS LOCATED ON THE CORNER OF SECOND AND MAIN STREETS.

business travelers." Other services, such as in-room Internet connections and extra phone lines, are standard features in many of the company's hotels.

However, some of Musselman Hotels' properties, such as the Holiday Inn Express in Cave City, are more resort oriented. The Cave City location is near several of Kentucky's most popular tourist attractions, including Mammoth Cave and Barren River Resort, and focuses primarily on the needs of vacationing families.

THE COURTYARD BY MARRIOTT

Musselman Hotels actively seeks out new locations in need of quality hotels. One of its newest hotels is right in its hometown of Louisville. The Courtyard by Marriott, located on South Second Street, opened its doors in June 1999 and became the first new downtown hotel in 10 years. "This is probably one of our best business facilities to date," says Kelley.

The new Courtyard by Marriott is located in the rejuvenated eastern segment of downtown. Its neighbors include Louisville Slugger Field, Waterfront Park, and, most important, the renovated and expanded Louisville Convention Center, which is only one block away. "Louisville is a real convention city, and many of our guests are business travelers who attend conventions," says Kelley.

To accommodate those travelers, the Musselman-owned Courtyard by Marriott is the only downtown hotel that provides direct, high-speed Internet access in every guest room. In addition, the hotel features spacious rooms and suites with a sitting area, a large work desk, and two telephones with voice mail and data ports.

Like all Musselman Hotels, the Courtyard by Marriott prides itself on providing outstanding service to all visitors, but guests are also likely to notice the unique design of the hotel. Musselman Hotels had to create a hotel that would fit into the vacant downtown location that was available. A beautiful addition to the downtown area, it boasts one of the best locations to view the Thunder Over Louisville fireworks and air show.

With its focus on outstanding customer service, Musselman Hotels has experienced phenomenal growth over the years. As it prepares for more growth in the future, the company will continue to focus on customer service and quality management at the corporate level and at each individual hotel.

Neace Lukens

While working in the Kentuckiana area, John Neace noticed that many talented local people were being pulled away from Louisville by larger, national insurance agencies. Wanting to keep the pool of qualified workers in Kentucky and southern Indiana, he formed the Neace Group in 1991. Starting with just six employees, he has seen his vision fulfilled and the company become a full-service brokerage and consulting firm that employs more than 280 highly skilled workers in 12 offices located throughout Kentucky, Indiana, and Ohio.

Neace Lukens has quickly grown to be the largest insurance agency based in Kentucky. Revenues have increased from $600,000 in 1991 to more than $16 million in 1999. "Our quality employees have been an essential part of our growth," says Neace. "In our business, knowledgeable, conscientious employees are absolutely a key asset."

Neace Lukens provides a full range of diversified services that include property and casualty insurance, risk management, employee benefits, third-party administration, and surety bonds for the firm's commercial lines of business. The company also offers a variety of personal insurance plans, as well as comprehensive estate planning.

Neace Lukens works with a variety of companies that are helping Louisville grow, including the contractor responsible for the construction of Louisville Slugger Field. Pictured here are (from left) John Neace, William Chilton, Richard Chilton, and Alan Jones (top).

Neace Lukens employs more than 280 highly skilled workers throughout Kentucky, Indiana, and Ohio (bottom).

Business Partners

Neace Lukens strives to establish long-term relationships with its growing list of clients, and to be there when its clients' needs change. Neace states, "We look at the total picture, from providing more coverage while reducing costs to constantly reviewing every client's needs to determine the best program for them. We want to be a part of the client's business." Through extensive research and continuous monitoring of the marketplace, the company offers custom-designed programs to ensure that its clients' needs are being met.

In its role as part of its clients' business teams, Neace Lukens offers true one-stop service for an employer's total insurance program. Neace Lukens provides and manages commercial insurance that includes property/casualty insurance, risk management services, and alternative market plans.

Moving far beyond typical commercial lines of business, Neace Lukens can also work hand in hand with an employer's human resources department to help manage employee benefits and find cost-effective alternatives for employee benefit plans. Under its employee benefits program, Neace Lukens provides and administers group life, group medical, dental, disability, cafeteria, self-insured, and qualified retirement plans.

Specialized Products

Neace Lukens has grown by providing the services needed by businesses in

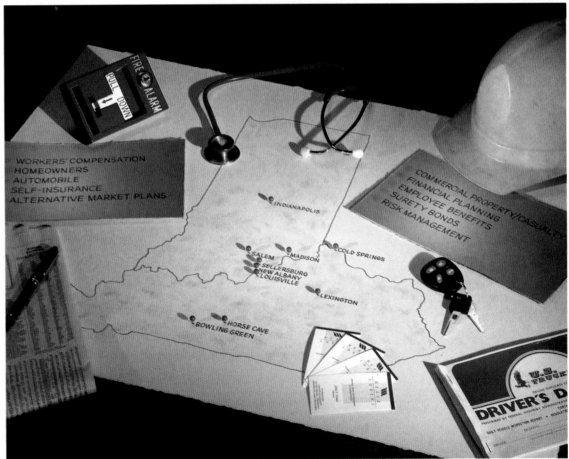

the communities it serves. This commitment has led the company's employees to focus their practice in various areas of specialty, including health care, surety, construction, and transportation. The company has created unique programs solely for the operation of these services. This niche marketing enables the company to offer services not readily available in the standard marketplace.

Surety bonds, essentially a guarantee backed by an insurance company, are a necessary commodity for almost every business. Neace Lukens' staff has more than 75 years of combined experience with surety bonds. The majority of surety bonding is done for the construction industry; surety bonds guarantee to owners that a building will be constructed to specifications, on time, and on budget. Because the construction industry requires fast-paced decision making, Neace Lukens provides a proactive approach with its clients by establishing surety relationships before a client actually needs them.

Neace Lukens also provides a variety of programs for the transportation industry. The firm designs and manages trucker liability, workers' compensation, cargo, umbrella, owner-operator, passenger, and fleet insurance programs. The experienced professionals at Neace Lukens understand the transportation industry, as well as their clients' unique needs. The staff has experience with the transportation industry, and understands that a fast-paced turnaround time is often critical. The staff also has experience in working with self-employed owner-operators, who have unique insurance needs. The company designs a customized program for each client.

Health care is another area in which Neace Lukens has distinguished itself from the crowd. Through its subsidiary Risk Management Services, a third-party administrator, the company provides numerous alternative and self-insured programs designed to help its clients manage risk. These programs have led to captive formation and numerous other cost-saving mechanisms for clients. Neace often states that people do not want to merely trade dollars with an insurance company. His company's innovative approaches have kept those dollars where they belong—in its clients' pockets.

Homegrown Success

Grown from a pool of talented employees in the region, Neace Lukens is dedicated to the Louisville area. While national insurance agencies are looking to consolidate their services and move out of the area, Neace Lukens plans to grow by adding more services and more offices throughout the region. Louisville is home to this company, and Kentucky is the focus of its plans for regional growth.

"As we grow, we want to keep our personal, homegrown atmosphere," says Neace. "Being geographically close to our clients helps us maintain the personal relationships that allow us to offer more creative and innovative approaches as insurance providers."

CLOCKWISE FROM TOP LEFT: NEACE LUKENS' SERVICE TEAM WORKS TO ENSURE CLIENTS' NEEDS ARE MET. THIS TEAM INCLUDES (FROM LEFT) SHARON ROSE, TERRY MCMAHEL, LARRY SCHAEFFER, AND CAROLYN STEWARD.

MARY JOE GRANTZ WORKS IN THE COMPANY'S INDIANA OFFICE.

NEACE LUKENS PROVIDES A VARIETY OF PROGRAMS FOR THE TRANSPORTATION INDUSTRY. THE FIRM DESIGNS AND MANAGES TRUCKER LIABILITY, WORKERS' COMPENSATION, CARGO, UMBRELLA, OWNER-OPERATOR, PASSENGER, AND FLEET INSURANCE PROGRAMS. PICTURED HERE ARE (FROM LEFT) NEACE, JONES, JEFF KAPPS, AND MARY CORBY.

1992–2000

1992 ADVENT

1992 Cambridge Construction Company

1992 OPM Services, Inc.

1993 Lightyear

1994 Adelphia Business Solutions

1997 Greater Louisville Inc.

1997 TechRepublic

1998 Muhammad Ali Center

1999 Insight Communications

ADVENT

"LOUISVILLE IS GOOD FOR BUSINESS," SAYS MARK SELLERS, A VETERAN of the environmental consulting and design industry and president of ADVENT. "From a progressive education system, to low crime and the quality of life that comes from a midsize city, to a probusiness culture in local government, Louisville's an ideal home for ADVENT." Since opening its Louisville office on 1992, ADVENT has become one of the area's premier professional engineering and environmental consulting firms, serving the top 10 percent of Fortune 500 companies in a variety of industries, including chemical distribution, synthetic organic chemical manufacturing, heavy manufacturing, and distilled spirits.

AT THE FOREFRONT OF ENVIRONMENTAL CONSULTING

One of ADVENT's key roles is to help companies understand various environmental regulations and how these rules will affect their operations. "Violations often occur because of companies' lack of awareness of new legislation, previously unknown problems, or a simple oversight," says Sellers. This is not surprising, he adds, considering that environmental regulations are constantly changing and are often set by various levels of state and federal government.

While ADVENT has found its niche in industrial consulting, the company's services go beyond environmental engineering issues. "We often must address the financial and regulatory aspects of these issues as well," says Larry Dietsch, executive vice president and regional manager.

"We work closely with our clients' management teams—including legal counsel, financial advisers, and technical staff—to bring our projects to a successful conclusion."

ADVENT is probably best known for services related to property acquisition and divestiture activity. "We have earned a name nationwide in the area of brownfield redevelopment work," explains Dietsch. "This is where a contaminated industrial site that would be difficult to develop receives special regulatory treatment, allowing for a risk-based solution to be proposed to the regulators. This way, the cleanup process is shortened dramatically, which, of course, greatly reduces costs. We have had significant input into these programs and are well known as the firm that knows what to do next."

THE HUMAN FACTOR IN A TECHNICAL FIELD

The key to doing the job right is individual service," says Dietsch. "Although it's a very technical field, there's a lot of one-on-one interaction with the client. In most cases, we function as an extension of the client's own environmental staff." The importance of close interaction has led ADVENT to be selective about its clients.

ADVENT WORKS CLOSELY WITH CLIENTS' MANAGEMENT TEAMS—INCLUDING LEGAL COUNSEL, FINANCIAL ADVISERS, AND TECHNICAL STAFF—TO BRING PROJECTS TO A SUCCESSFUL CONCLUSION.

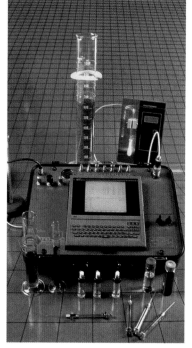

ADVENT has attracted professionals who not only are highly experienced in environmental consulting, but also have in-depth experience in the industries they serve. "Most of our senior staff are registered professionals," says Rich Coomes, principal and manager of ADVENT's Louisville operations. "Our clients are confident that their consultant truly understands their current needs and the trends affecting them."

Diversification Fuels Growth

ADVENT has been growing steadily since opening its doors. Beginning with only one office and two employees, the firm today has a diverse staff of professionals and offices located in Louisville; Charlotte; St. Louis; Charleston, South Carolina; and Barcelona, Spain.

In the beginning, ADVENT concentrated on consulting engineering for the chemical industry. By 1995, the firm had begun to diversify. The company added international and government service operations, focusing on a number of contracts available in Spain and Portugal, as well as those available from the Department of Defense. The diversification continued with ADVENT's certification as a small disadvantaged business (SDB) in 1999, and with the addition of professionals focused on remediation and unexploded ordnance services.

Firmly Attached to Louisville

ADVENT's employees in Louisville say they are delighted with the city for many reasons. According to Sellers, one of the reasons the firm chose to locate in Louisville is local government officials' probusiness attitude. Sellers cites major local activities—such as airport expansion, downtown development along the riverfront, and recognition of the need to redevelop older industrial properties—as examples of the proactive political climate in the city. Plus, Louisville is located near many of the types of industries that make up ADVENT's client base.

With the rapid advancements of technology, new contaminants produce new regulations—regulations that mean a continuing need for assistance to find the acceptable balance between efficient operations and environmental responsibility. Offering expertise in the field of engineering and environmental consulting, ADVENT stands ready to help companies meet this ongoing challenge.

ADVENT TRULY UNDERSTANDS ITS CLIENTS' NEEDS AND THE TRENDS AFFECTING THEM.

Cambridge Construction Company

When Larry Heck and his partner Barry Koenemann decided to start their own construction firm in 1992, the Louisville economy was not experiencing the building boom it is undergoing today. "When we started business, we were willing to build almost anything for a buck," says Heck, who today serves as president of Cambridge Construction Company. The firm survived that first year with total billings of less than a million dollars. By 1999, the company had grown to posting sales of more than $57 million.

Cambridge Construction Company was established when the two partners purchased the former Louisville regional office of the R.W. Murray Company. Since then, with its sister company United Construction Enterprises Co., the firm has operated offices in Louisville, St. Louis, and Indianapolis. The companies are currently providing construction services in six states.

Cambridge specializes in design/build commercial, industrial, and retail projects, including office buildings, retail shopping centers, banks, medical office and treatment facilities, schools, and churches. A major component of the company's current business is large industrial distribution centers with more than 4.5 million square feet of space, completed in the past four years. Using tilt-wall and precast concrete construction wall panel systems, Cambridge has completed distribution centers up to 523,000 square feet with 40-foot clear heights.

One-Source Construction Provider

Our goal is to provide our clients with a single source of responsibility for all design disciplines, as well as the construction of the project," says Heck. "We help our clients with every aspect of the construction process. We get them off the ground by helping them with zoning issues; then, we work to get the utilities and other infrastructure in place. During the planning process, we assist clients in determining the most cost-effective design for their building and the best land usage."

With a total of 60 employees, Cambridge staff professionals begin the design/build process by providing in-house layout and design, monitoring soils testing, and determining material selections and specifications. Final construction documents are prepared by outside professionals selected from a list of design firms for their expertise in the area of work required by each specific project. All of this is done in close association with the client to ensure the highest level of customer satisfaction.

Although a design/build specialist, Cambridge is also a valuable project team member with designers contracted directly by the firm's clients. Cambridge provides value engineering that results in using more efficient construction methods and the latest technology available to provide the maximum value from every construction dollar.

Client Relations

One of the keys to Cambridge's success is the long-term relationships

A MAJOR COMPONENT OF CAMBRIDGE CONSTRUCTION COMPANY'S CURRENT BUSINESS IS LARGE INDUSTRIAL DISTRIBUTION CENTERS WITH MORE THAN 4.5 MILLION SQUARE FEET OF SPACE, COMPLETED IN THE PAST FOUR YEARS. USING TILT-WALL AND PRECAST CONCRETE CONSTRUCTION WALL PANEL SYSTEMS, CAMBRIDGE HAS COMPLETED DISTRIBUTION CENTERS UP TO 523,000 SQUARE FEET WITH 40-FOOT CLEAR HEIGHTS.

the firm has built with its clients. The company's client list includes some of the largest real estate developers in the nation, as well as a number of Louisville's most prestigious companies.

Cambridge has buildings in all of Louisville's major business and industrial parks, including Riverport, Eastpoint Business Centre, Louisville Industrial Center, Bluegrass Industrial Park, Commerce Crossings, Clark Maritime Center, and Hurstbourne Green. The firm has also constructed some of the most modern distribution facilities in the area, including projects for GEA Parts Distribution, Genecom, Master Lock, Campbell Group, R.R. Donnelley & Sons, Exel Logistics, Brown & Williamson Tobacco, JB Hunt Transportation, Sebastian International, Hilliard Lyons, Unidial, and Thomas Industries.

Since its founding, Cambridge Construction Company has been striving to be at the forefront of the construction industry, fostering long-term client relationships and working as a full-service partner with its clients from a project's beginning to its end. As Cambridge continues to grow, the firm's dedication to this high level of customer service will ensure its success for decades to come.

CAMBRIDGE SPECIALIZES IN DESIGN/ BUILD COMMERCIAL, INDUSTRIAL, AND RETAIL PROJECTS, INCLUDING OFFICE BUILDINGS, RETAIL SHOPPING CENTERS, BANKS, MEDICAL OFFICE AND TREATMENT FACILITIES, SCHOOLS, AND CHURCHES.

OPM SERVICES, INC.

GROWTH AND CHANGE HAVE CHARACTERIZED LOUISVILLE'S RECENT history. While everyone seems to be enjoying this growth, a few companies are ensuring that the growth and quality of life will continue far into the new century. One of those is OPM Services, Inc., a company dedicated to helping young businesses grow and prosper. ■ OPM Services was formed in 1992 to provide management, support, and funding to entrepreneurial start-ups. "Louisville is becoming a hotbed for entrepreneurial companies, and OPM Services was created to be the catalyst for new businesses," says founder Kent Oyler.

OPM Services has succeeded in creating an extended family of companies that have been enhanced by its expertise. These young companies received help with the myriad details entrepreneurs find necessary to start a business. "We provide access to start-up capital, business expertise, and a drive to make things happen," says Oyler.

OPM strives to offer just what each young company needs. For some companies that means a partner; for others, it means management expertise. Other new ventures simply rely on OPM's back-office support during the start-up stage. Whatever it takes, OPM primes the fledgling companies for growth.

VARIED FIELDS, VERITABLE SUCCESS

OPM Services' first venture into the entrepreneurial start-up field was Icelease, a company that serves as a lessor of high-capacity ice-making equipment. Just a few short years after its founding, Icelease—with aid from OPM Services—purchased the Tube Ice and Turbo Refrigeration divisions of the Henry Vogt Machine Company. Today, Icelease is the world's largest manufacturer and lessor of high-capacity ice-making equipment.

Another company started under the direction of OPM Services during the 1990s is doing very well. Power Flats, Ltd. is an owner and operator of heavy-duty flat railcars. Power Flats has expanded its business by operating railcars for several car owners—in addition to its own—under the LNAL marking. OPM Services helped it achieve a leading place in the market by providing a combination of highly competent back-office services management and financial backing. Power Flats is now the nation's largest operator of heavy-duty flatcars.

Founded in 1997, UniStar LLC is a Louisville-based company competing in the exploding field of electronic business-to-business purchasing management. OPM Services' support during UniStar's start-up helped it capture national accounts with the National Thoroughbred Racing Association, Caesar's World, and numerous other companies.

EXPANSION INTO NEW TECHNOLOGIES

In 1997, OPM Services began executing opportunities in the telecommunications and technology sectors. Oyler and then-business partner David Gibbs founded CATV.net, LLC. Shortly after its founding, CATV.net merged with a Denver-based start-up to form High Speed Access Corporation (HSAC), a business dedicated to bringing broadband Internet access to ex-urban markets, primarily through partnerships with cable television system owners. After only 14

THE NEW HSA BROADBAND BUILDING FOR KENTUCKY PUBLIC RADIO WAS FUNDED IN PART BY A DONATION FROM OPM SERVICES, INC.

months of operation, the company went public with the largest initial stock offering ever for a Kentucky-based company. HSAC (Nasdaq: HSAC) has enjoyed a market value in excess of $1 billion, and has already experienced its first spin-off, Darwin Networks, Inc. Darwin provides high-speed Internet access and data solutions to businesses, apartments, hotels, and college dormitories using wireless and digital subscriber line (DSL) technologies. "Darwin's sponsor was Chrysalis Ventures, who saw an opportunity to expand broadband Internet service into new areas," says Oyler.

The talent and expertise developed helping diverse start-ups has now been put behind bCatalyst, a business accelerator founded by principals from OPM and Louisville venture capital firms. OPM provides the back-office support as bCatalyst slingshots promising start-ups to top-tier venture capital funding and beyond.

Success in business brings with it civic obligations. OPM's owners and financial partners have stepped up to, and raised, the bar. The privately funded, $5 million HSAC/Darwin New Business Challenge with Metro United Way and the new HSA Broadband Building for Kentucky Public Radio are two highly visible examples of community support.

As the businesses touched by OPM Services grow, Oyler hopes they, too, will be inspired to assist other new companies. "I believe Louisville is one of the best possible places to live and to start a company. I am dedicated to helping start-up enterprises here because I love the planning and the excitement involved. Right now is such an exciting time to be a Louisvillian."

VOGT ICE IS THE WORLD'S LARGEST MANUFACTURER AND LESSOR OF HIGH-CAPACITY ICE MACHINES.

HSAC'S NETWORK OPERATIONS CENTER IN LOUISVILLE MONITORS BROADBAND CONNECTIVITY NATIONWIDE.

LIGHTYEAR

HELPING CUSTOMERS BUILD THEIR BUSINESSES THROUGH THE USE OF leading-edge, converged communications services—that's the mission of Lightyear, an integrated communications provider, offering businesses everything from advanced data services such as frame relay and point-to-point to nationwide high-speed Internet access, complete voice services, and multimedia conference services.

THE LIGHTYEAR REVOLUTION

During the spring of 2000, the company—which was founded in 1993 under the name UniDial—began the transformation to Lightyear. "With a solid history as UniDial, we feel the name Lightyear reflects where we are heading," says Senior Vice President G. Henry Hunt. "We are revolutionizing ourselves and, we believe, the industry as a whole with this change."

Lightyear is changing the way medium-size businesses utilize communication technology, as well as the way communication networks are built and operated, through its own nationwide network. This network brings the latest in technology to customers, which allows for more efficient bandwidth allocation and greater control of costs.

Lightyear's new network, which is available exclusively to business customers in major markets throughout the United States, features cutting-edge technology that integrates data, Internet, and voice onto a single, packetized circuit. This allows the business customer to utilize a single T-1 line—instead of the two to three needed for traditional technology. Lightyear supplies an integrated access device (IAD) that handles the converged signal at each customer's premises and allows the Lightyear Network Operations Center (NOC) to be directly connected to the customer's point of business. Repairs to the customer's system can be made remotely from the NOC in Louisville with just the touch of a button. The benefits to the customer include better service, greater utilization of bandwidth, lower equipment costs, and substantially lower monthly access charges.

"This network helps us help our customers stay light-years ahead of their competition," says Mike Johnson, chief technology officer for Lightyear. "We're all about helping our customers take advantage of the efficiencies of the new economy. To do that, you need bandwidth, used as efficiently as possible. We really do believe our network is unique, and our products are designed to deliver the kind of capacity and connections that will make that all possible. Because WorldCom and Williams Communications are part owners of Lightyear, we are able to bring access to customers anywhere, anytime."

J. SHERMAN HENDERSON III IS PRESIDENT AND CEO OF LIGHTYEAR.

LIGHTYEAR'S EXPLOSIVE GROWTH HAS INCLUDED THE CONSTRUCTION OF A 75,000-SQUARE-FOOT FACILITY IN LOUISVILLE'S EASTPOINT BUSINESS CENTER.

LIGHTYEAR—FORMERLY KNOWN AS UNIDIAL—WAS BORN ON A NAPKIN AT A DENNY'S RESTAURANT IN LOUISVILLE. THAT INFAMOUS MOMENT REMAINS A PART OF THE LIGHTYEAR CULTURE, SO MUCH SO THAT AT THE COMPANY'S FIFTH ANNIVERSARY PARTY, HENDERSON WORE A DENNY'S UNIFORM AND PASSED OUT MEALS CATERED BY DENNY'S TO THE LIGHTYEAR EMPLOYEES.

A Growing Presence on the Louisville Business Scene

With a 20 to 60 percent growth rate every year since its inception, Lightyear's numbers are impressive. In 1999, the company's annual revenue reached $213 million. As of April 2000, Lightyear had more than 700 employees, with 400 based in Louisville; four headquarters offices in Louisville; and 20 offices nationwide, where its products are sold through an army of salespeople and hundreds of independent authorized agents to more than 300,000 customers.

Lightyear has grown by 100 employees every year since its founding, and this trend is expected to continue. While much of that growth has occurred in the national sales force, engineering, and information technology (IT) areas, substantial growth has also taken place in the hub of Lightyear's operation: its call center at the Eastpoint Business Center. Here, thousands of calls a day pour in from customers, whose needs are met by highly trained service personnel who view all customer account and service status information through a single system. The company's NOC, where the network is continuously monitored and new products are tested, is located in its own building on Linn Station Road. Lightyear also has two locations in the Hurstbourne Green Office Complex for corporate personnel and the Louisville sales and marketing staff.

A Rich History

The national company began on the back of a napkin at a Denny's Restaurant in Louisville. J. Sherman Henderson III, Lightyear's president and CEO, had meeting after meeting with potential employees and allies there, armed only with a good idea, a handful of staff, and an innovative agent program.

"I had already built and sold another small telecom firm, Charter Communications, but this time I wanted to do it better," Henderson says. "We started this company by developing solid relationships with vendors, like WilTel, that allowed us to resell their long-distance products. Then, we recruited the very best independent authorized agents to sell our products. It wasn't easy at first, but once we got the ball rolling, it was incredible what we could do. We really helped kick open the doors to true competition in telecom."

UniDial quickly became known as one of the fastest-growing enterprises in the industry, and Henderson was repeatedly named to *Phone+* magazine's 25 Most Influential People in Telecom list. Soon, the company's network of agents spanned the country from coast to coast, as did its sales force. UniDial began building its back-office operation to include a 24-hour, seven-day-a-week call center, and became well known as having the best training, support, and products to sell. The Telecom Act of 1996, which legislated more open markets for competitive providers like UniDial, brought new opportunities for a host of other products like local phone service, data networking, enhanced calling card services, conferencing, and much more.

As UniDial grew, so did the size of its allies. WorldCom bought an 8 percent share in the company. Williams bought a 10 percent stake. Bell Atlantic signed a $300 million contract allowing UniDial to resell Bell Atlantic local phone service in 14 northeastern states—the largest resale agreement in Bell Atlantic's history. Clearly, UniDial was on the track to becoming one of Louisville's leading corporations.

Henderson is a big believer in Louisville as a place to build both a family and a business. He remains an active member of the local business community. "A lot of people think Louisville is not a tech-savvy town, but that is changing fast," Henderson says. "It's exciting to see so many tech companies sprouting up downtown and along the Shelbyville Technology Corridor. I'm proud Lightyear has been a part of it. It just goes to show you—you can accomplish anything if you just believe. Lightyear is living proof of that."

Adelphia Business Solutions

Louisville has been a source of tremendous growth for Adelphia Business Solutions since 1994. Adelphia Business Solutions is a national integrated communications provider that offers an extensive suite of services to businesses, including local voice, long-distance, messaging, Internet, and enhanced data services. Adelphia Business Solutions is making its mark throughout the United States by establishing fully integrated communications services in major metropolitan areas that previously had to rely on a limited number of telecommunications providers.

Entering a market with existing telecommunications providers presents a huge challenge for any competitor. Adelphia Business Solutions General Manager Paul Carlisle addresses that issue by rallying his sales forces with the words he lives by: "Challenge means opportunity." Those are clearly words that ring true. Louisville is a leader in overall market sales nationally for the company.

Backed by Nearly 50 Years of Experience

Adelphia Business Solutions is part of Adelphia Communications Corporation, a company with nearly 50 years of experience in the communications industry. Adelphia Communications, one of the largest cable television companies in the United States, was founded in Coudersport, Pennsylvania, by two brothers, John and Gus Rigas, in 1952 as a cable television franchise. The brothers went door-to-door trying to convince neighbors to take down their television antennas in exchange for more channels via cable technology. By 2000, more than 5 million people across the country had subscribed to Adelphia Communications' cable TV services.

In 1991, Adelphia Communications formed a company known as Hyperion Communications to enter into the telecommunications industry. In 1999, Hyperion changed its name to Adelphia Business Solutions to better reflect its parent company and to establish one powerful brand in the telecommunications industry. The core of the company's business has been building state-of-the-art fiber-optic networks in metropolitan business markets.

Since its inception, Adelphia Business Solutions has been dedicated to delivering the best. From the beginning, it built its fiber-optic networks with the most advanced materials; today, its networks are enhanced with state-of-the-art Lucent 5ESS switches to deliver the highest-quality communication. Technology enhancements continue to be an important part of Adelphia Business Solutions' strategy for delivering quality service.

Adelphia Business Solutions closed the year 1999 with revenues of more than $154 million and more than 6,700 local route miles of fiber installed and connected to more than 2,000 buildings. Throughout 2000, Adelphia Business Solutions will expand into the western United States, and by the end of 2001, will serve more than 200 cities throughout the nation on its fully redundant, 30,000-mile local and long-haul fiber-optic network.

Services for Every Business

In addition to its fiber-optic network, the company provides local and long-distance services specifically

Paul Carlisle is the general manager of Adelphia Business Solutions.

designed to fulfill the diversified communication needs of today's business community. Adelphia Business Solutions provides analog business lines, PBX trunking, and sophisticated ISDN technology. The long-distance service complements local voice service, providing long-distance and international communication service with costs billed in six-second increments to ensure accurate charges.

Adelphia Business Solutions also offers a customized phone system for businesses with multiple sites. Ordinarily, companies with multiple sites throughout a city have to dial a seven- or 10-digit number to call from one location to another. Local charges apply to these calls and cost businesses money. Instead, Adelphia Business Solutions can create a system that allows employees to dial between company sites using only a four-digit extension, eliminating the charges typically associated with these calls. The firm can also provide customized messaging services, including voice mail, fax mail, and automated attendant.

Another aspect of Adelphia Business Solutions' total communications package is its business Internet services—services specifically designed to give businesses quick, productive, and cost-effective Internet access. Adelphia Business Solutions can provide businesses with a dial-up or dedicated Internet connection, or serve as the host for a business' Web page. The company also offers dedicated Internet services that deliver direct, high-speed connections ideal for businesses with sophisticated Internet requirements.

GENUINE RELATIONSHIPS

More than the quality services the company offers, Carlisle attributes the success of Adelphia Business Solutions in Louisville to the hard work ethic he insists on from himself and his employees. "Good, old-fashioned relationship building is another element to my formula for success," says Carlisle. "A good reputation in the business community goes a long way in winning over clients."

That kind of philosophy translates into business relationships with numerous high-profile clients such as Humana Inc. and PNC Bank. Carlisle has led his team to Kentucky companies and organizations large and small by delivering services with a personal touch.

"As an integrated communications provider, we can offer exceptional quality with the most advanced technology at a very competitive price," says Carlisle. "I believe that people want to do business with Adelphia Business Solutions because of our commitment to genuine relationships based on mutual respect that transcends traditional customer service relations."

GIVING BACK TO THE COMMUNITY

Adelphia Business Solutions' commitment to the Louisville area is evident in its support of community outreach sponsorships. The University of Louisville Cardinal Park development is one such project that has received corporate funding from the company, totaling $1.75 million. A generous contribution for the construction of the recently completed Louisville Slugger Stadium, home of the RiverBats, is another example of the kind of support Adelphia Business Solutions lends to Louisville.

"As a busy father of two children and a youth soccer coach, the sponsorships are near and dear to my heart. Additionally, lending our advanced technology capabilities to the Louisville community will help Louisville continue to realize world-class-city status," says Carlisle. With focused solutions and dedication to the community, Adelphia Business Solutions gives businesses and the community the power they need to succeed.

ADELPHIA BUSINESS SOLUTIONS' COMMITMENT TO THE LOUISVILLE AREA IS EVIDENT IN ITS SUPPORT OF COMMUNITY OUTREACH SPONSORSHIPS. THE UNIVERSITY OF LOUISVILLE CARDINAL PARK DEVELOPMENT IS ONE SUCH PROJECT THAT HAS RECEIVED CORPORATE FUNDING FROM THE COMPANY, TOTALING $1.75 MILLION.

ADELPHIA BUSINESS SOLUTIONS' OFFICES ARE LOCATED IN A RESTORED HISTORIC BUILDING IN DOWNTOWN LOUISVILLE.

GREATER LOUISVILLE INC.

During 1997, a group of Louisville-area business leaders developed a new vision for the community's economic future. That vision called for Louisville to be transformed into one of America's economic hotspots—a city characterized by increasing population, more high-growth entrepreneurial companies, and greater opportunities for highly skilled, highly paid technical and professional workers. Part of their vision included a new model for a combined economic development group and chamber of commerce. The end result was the formation of Greater Louisville Inc.-The Metro Chamber of Commerce (GLI).

GLI is a business leadership organization dedicated to helping its members and the community grow and prosper. The corporation serves as the region's primary economic development organization, with responsibility for business recruitment, expansion, and entrepreneurship. In addition, GLI provides its members with such services as professional development, government advocacy, research assistance, workforce development, minority business development, and networking opportunities.

THE LOUISVILLE MEDICAL CENTER IS BECOMING A HUB OF TOP RESEARCH AND DEVELOPMENT (TOP).

GREATER LOUISVILLE INC.-THE METRO CHAMBER OF COMMERCE (GLI) MAKES ITS HOME IN THE HISTORIC DISTRICT OF WEST MAIN STREET (BOTTOM).

ECONOMIC DEVELOPMENT

The mission of GLI's economic development department is to provide regional solutions for global growth opportunities in the Greater Louisville metropolitan area. As Louisville's lead organization for business attraction, GLI coordinates location and site selection for clients and serves as a liaison between businesses and key individuals in the public and private sectors.

GLI has done a stellar job: Successes in 1998 included the nation's largest economic development project (a billion-dollar expansion of the UPS international air hub in Louisville). GLI followed this achievement with a record year in 1999. Its economic development efforts helped 51 companies expand or locate in the region, creating 5,645 new jobs—at an average salary of $39,257—and more than $387 million in new capital investment.

INNOVATIVE WORKFORCE SOLUTIONS

Right now one of our top priorities is to pursue ways to find and keep the kind of skilled workforce that Louisville will need to support continued economic growth," says Steve Higdon, president and CEO. GLI has been hard at work creating innovative workforce solutions, including one of the nation's most creative solutions to workforce development, Metropolitan College, a partnership of UPS, the State of Kentucky, and three Louisville-area colleges. Students who attend Metropolitan College receive college tuition in addition to their UPS salary for a part-time night shift. Annually, more than 1,000 students enroll in Metropolitan College, which has received favorable national attention in publications such as *The Wall Street Journal*, *The New York Times*, *The Washington Post*, and *U.S. News & World Report*.

GLI also offers employers one-stop recruiting assistance, including access to a pool of qualified applicants; referrals, screening, and assessments; opportunities for participation in job fairs; labor market information; facilities for on-site interviews; outplacement services; and disability awareness training. GLI has also developed an economic model to help employers calculate the cost of employee turnover. Other programs to help create a larger workforce of high-quality employees include the Institute of Care Management, a communitywide consortium of public and private university nursing programs addressing the needs of the growing managed-care profession.

GLI'S EFFORTS ARE TARGETED AT TRANSFORMING THE CITY INTO ONE OF AMERICA'S ECONOMIC HOTSPOTS.

SERVICES FOR THE ENTIRE BUSINESS COMMUNITY

GLI takes its business outreach to the company's doorstep, making calls and talking to local business owners. In 1998, the city and county contracted with GLI to work with existing businesses in the community and help them expand. The organization makes sales calls to local businesses and assists them by determining programs and tax incentives the companies might be eligible to receive—all free of charge.

Young companies trying to grow in Louisville receive extra attention through The Enterprise Corporation, the entrepreneurial development arm of GLI. The Enterprise Corporation is dedicated to providing a central source of information, advice, and contacts for entrepreneurs. By providing CEO forums, business plan peer reviews—called Ugly Baby Forums—and a number of networking and informational opportunities, The Enterprise Corporation hopes to spur the future of homegrown success stories.

For its members, GLI offers a wide variety of services, including Business at Breakfast meetings, featuring experts from around the country; small-business development workshops; CEO roundtables; and cost-saving programs for long-distance calling, credit card processing, and Internet service. GLI is striving to ensure that all Louisville-area businesses get the business solutions they need to work.

EYE ON THE FUTURE

Moving beyond the current business climate, GLI keeps its eye on the future and strives to make sure the Louisville economy will continue to run strong. At the start of the 21st century, getting Louisville connected to the Internet is one of GLI's top priorities, as it works to ensure that area businesses will be able to compete globally.

GLI has also focused growth efforts on two targeted business industries: logistics and biomedicine. Both of these areas are already well represented in Louisville, with the UPS hub providing services for logistics and world-renowned medical centers attracting new talent every day.

However, the community will have to continue its innovative approaches to economic growth if the vision set for Louisville by its business leaders is to become a reality. "There's a lot of work to do," says Higdon. "But we've got great people and a great product. The future has never looked brighter for our community."

SITUATED ON THE OHIO RIVER, LOUISVILLE IS A THRIVING CENTER FOR RIVER COMMERCE, IN ADDITION TO BEING EASILY ACCESSIBLE BY LAND, AIR, AND RAIL.

GLI'S ANNUAL SHOWCASE GREATER LOUISVILLE IS THE REGION'S PREMIER BUSINESS-TO-BUSINESS TRADE SHOW.

TechRepublic

TechRepublic is revolutionizing the relationship between buyers and sellers of technology. Through this premier Web site, information technology (IT) professionals connect with peers and vendors through content, products, and services that enable collaboration, career development, and commerce. Members can create professional profiles, seek advice from their peers, sign up for E-mail subscriptions on more than 40 IT-related topics, find jobs, and keep abreast of upcoming technology events and conferences. TechRepublic is organized into Republics, divided by job category. In addition to the Republics, the site also hosts TechProGuild, a premium on-line service, featuring in-depth technical content and the Enterprise Resource Planning (ERP) site.

"In less than a year, TechRepublic has gone from a little-known garage band from Louisville, Kentucky, to playing giant stadiums of IT pros," says Tom Cottingham, CEO. "We've gone mainstream and, today, are an integral part of the IT world."

When Cottingham founded the company in 1997, he envisioned a community where IT professionals could learn about the industry's best practices and interact with their peers to share problems, ideas, and solutions. Cottingham had already spearheaded numerous IT successes, including cofounding the Cobb Group, which produced technology-focused newsletters. For two years, Cottingham served as president of the Cobb Group. After Ziff-Davis acquired the company in 1991, Cottingham became a founding board member for *Family PC* magazine and the lead executive of *Computer Gaming World*, both owned by Ziff-Davis.

While Cottingham provided TechRepublic with its vision, cofounder Kim Spalding provided the financial savvy. Spalding had served for seven years as the vice president of finance and administration at the Cobb Group after Ziff-Davis acquired the company. She also was the business manager for *Computer Gaming World*. Together, Cottingham and Spalding assembled a team of seasoned professionals from the IT publishing, marketing, and operations fields.

"We asked tech professionals what they wanted, and their response, across the board, was to be able to talk and share advice with their peers," Cottingham says. "So, that's what we created. TechRepublic is a virtual community where IT professionals can learn best practices, share problems, and create solutions together."

In March 2000, the start-up was purchased by the Gartner Group, a world leader in providing technology research, consumer and market intelligence, consulting, conferences, and decision-making tools. TechRepublic remains an independent subsidiary of Gartner.

"Success in this business is defined by content, brand, and reach," says Michael Fleisher, president and CEO of Gartner. "The combination of Gartner's authority and TechRepublic's community access addresses the needs of the total group of IT influencers and professionals."

The Spirit of the Internet

TechRepublic believes in the creative, liberating spirit of the Internet, and the company's work environment reflects that belief. Employees love their work, and newcomers like Heather Virga find their outlook contagious. Virga joined TechRepublic as project manager with the California-based sales and marketing team.

"I looked at the Web site and knew this company was positioned to become the best IT site on the Web," Virga says. "After two interviews with the marketing team, I became enthralled with the group's energy and enthusiasm. I knew this was a company I wanted to be a part of."

Staff members are given the space—literally and figuratively—to think creatively. Employees are provided with offices rather than

> TECHREPUBLIC IS REVOLUTIONIZING THE RELATIONSHIP BETWEEN BUYERS AND SELLERS OF TECHNOLOGY. THROUGH THIS PREMIER WEB SITE, INFORMATION TECHNOLOGY (IT) PROFESSIONALS CONNECT WITH PEERS AND VENDORS THROUGH CONTENT, PRODUCTS, AND SERVICES THAT ENABLE COLLABORATION, CAREER DEVELOPMENT, AND COMMERCE.

"In less than a year, TechRepublic has gone from a little-known garage band from Louisville, Kentucky, to playing giant stadiums of IT pros," says Tom Cottingham, CEO. "We've gone mainstream and, today, are an integral part of the IT world (top left)."

Jennifer Recktenwald and Sarah Stom review notes from a customer roundtable (top right).

Bill Johnston (bottom left) and Greg Gorman (bottom right) enjoy TechRepublic's casual work environment.

cubicles, and are entrusted to manage their own workloads without constant supervision.

"We hire creative, productive people, so it only makes sense to give them the freedom and privacy they need to do their jobs," Spalding says. "They need quiet. They need to be able to think things out."

Employees are accustomed to expressing their ideas and opinions. Just ask Mike Jackman. When he decided to climb 20,285-foot Imja Tse in Nepal, he thought it would be fun to bring along some high-tech gear and see how it performed in extreme conditions. TechRepublic agreed. The company sponsored the expedition and covered Jackman's three-week adventure on-line.

TechRepublic employees can participate in on-site tai chi classes three times a week or blow off steam with a quick game of Ping-Pong. The company also hosts numerous special events, including a party at the local comedy club, a workday break for the premiere of *Star Wars: Episode I-The Phantom Menace*, and a mimosa brunch to celebrate Gartner's acquisition of TechRepublic.

TechRepublic also encourages community involvement. The company sponsors an aggressive internship program that recruits college students, providing them with experience in the high-tech field. Each division of the company participates in the program. Additionally, the company provides an extra personal day to employees who volunteer with certain local charities.

The Future

Very quickly, TechRepublic has moved from a struggling start-up to a major contender in the IT community. The company is poised to capitalize on the growing business-to-business Internet economy by virtue of the open forum it provides for vendors, retailers, buyers, and users to discuss products and issues.

That's what attracted Lisa Kiava to TechRepublic in 1999. Kiava previously worked as a television news reporter in Louisville, which is one of the country's top local news markets. As an editor for TechRepublic, she is given the time and space to explore issues in the IT industry.

"I was intrigued with the opportunity to work for an Internet start-up," Kiava says. "I knew there was risk involved. But because the company was growing so quickly, I knew it offered exceptional opportunities for advancement. As a TV news reporter, it's exciting to cover a variety of topics, yet it's frustrating to write in a very general way about a subject. TechRepublic covers a wide variety of business and IT topics, but allows me to write and research these topics in depth."

"Everyone here realizes the vital part we are playing in making the Internet a more useful tool," Cottingham says. "What we do at the office is changing the future as we help shape the Web."

Muhammad Ali Center

The Muhammad Ali Center—scheduled to open in 2003—is not intended to glorify the person for whom it is named, but to illustrate how the boxing legend has conducted his life and found the courage to stand up for his beliefs. The Center, based in the city where Ali was born, looks at the legacy Ali has built for himself. ■ Located on downtown riverfront property donated by the city, the Center will house various exhibits and activities, serve as a museum, showcase Ali's boxing career, and promote the ideals of peace, racial harmony, and respect that Ali holds dear. The torch Ali used to light the flame at the 1996 Olympic Games in Atlanta will be prominently displayed within the Center.

Ali's universal appeal is expected to attract more than 400,000 visitors a year and generate millions of dollars in associated revenue, creating more than 350 jobs within the Louisville area. The Center also anticipates attracting a worldwide audience through Internet-based programs, lectures, and school curriculum packages.

Living the Ideals

The Center is not just a museum built to honor Ali, but a way to carry out the mission of his life. The excitement of Ali's life will be captured through state-of-the-art exhibits, multimedia programs, hands-on learning, and opportunities for personal reflection. Part of the mission of the Center is to preserve and share Ali's legacy and ideals.

Through the Center's appeal to the heart, spirit, and imagination, children and adults will actively engage in making commitments to personal growth, discipline, tolerance, and respect. This will allow those who participate to realize their dreams and use their full potential to work toward them.

The Muhammad Ali Institute will be developed by the Center in conjunction with the University of Louisville. The Institute will focus on peacemaking, conflict resolution, and dispute mediation, and will bring together scholars, activists, educators, and other individuals from around the world. Those who come together through the Institute will advance their work in areas of human rights, alleviation of world hunger, physical and spiritual wellness, and sports ethics.

The Experience

The journey through the Center will begin when visitors experience Ali's life through a multimedia presentation. The trip through the Center will allow guests to "become" the

THE MUHAMMAD ALI CENTER, WHICH WILL BE LOCATED ON DOWNTOWN RIVERFRONT PROPERTY DONATED BY THE CITY, WILL HOUSE VARIOUS EXHIBITS AND ACTIVITIES, SERVE AS A MUSEUM, SHOWCASE ALI'S BOXING CAREER, AND PROMOTE THE IDEALS OF PEACE, RACIAL HARMONY, AND RESPECT THAT ALI HOLDS DEAR.

boxing celebrity, and travel through his life and across the globe as a humanitarian and goodwill ambassador.

The Center will show how Ali gained the respect of world leaders by using his fame as a tool for fighting poverty, hunger, and intolerance. Visitors to the Center will also gain an understanding of how life can be made better through acts of charity, respect, and tolerance. The World Ambassador's Hall of Fame will allow visitors to learn about other humanitarians as well.

Center visitors will get a behind-the-scenes look at Ali's progress toward becoming a three-time winner of the Heavyweight Champion of the World title, and will also have a chance to enter a gymlike space to test their own physical abilities. Visitors will be able to step into the world of professional boxing and see how important family, friends, and fans were to Ali in the building of his values and success.

Support from the Community

Plans for the $80 million Center were unveiled in 1998, and quiet fund-raising promptly began. Local, national, and international fund-raising efforts were put into place to obtain the necessary capital. The largest contribution was $10 million, approved by the Kentucky General Assembly, while other donations were given anonymously. The two largest civic foundations in the city—the James Graham Brown Foundation and the Gheens Foundation—also added their support in 2000 by awarding major grants for the development and endowment of the Center, joining ranks with the W.L. Lyons Foundation, which awarded a grant to the Center in 1999.

The Center enjoys enormous support from local and national leaders. Kentucky Governor Paul Patton, the General Assembly of Kentucky, Louisville Mayor David Armstrong, and Jefferson County Judge/Executive Rebecca Jackson are among those on the list of local supporters.

Members of the national advisory board for the Center include Seth Abraham of HBO Sports; Bob Costas and Dick Ebersole of NBC Sports; Sean McManus of CBS Sports; General Colin Powell, retired; Diane Sawyer of ABC News; former New York Governor Mario Cuomo; Dr. Vartan Gregorian of the Carnegie Corporation; and actors/comedians Billy Crystal and Robin Williams.

The Muhammad Ali Center is well on its way from being a dream to becoming a reality. With Ali as their guide—and as the voice of the Center—visitors are sure to leave with a deeper understanding of his life experiences, as well as a new perspective of their own.

THE MUHAMMAD ALI CENTER IS SCHEDULED TO OPEN IN 2003.

Insight Communications

Customers of Insight Communications have come to expect a lot more from this company than basic cable service. One of the nation's largest cable operators, Insight offers its customers a variety of state-of-the-art entertainment, educational, and information services in each of the many communities it serves. Headquartered in New York City, Insight has operations in Kentucky—where it is the largest cable operator—as well as in Illinois, Indiana, Ohio, and Georgia.

The company's primary business is the distribution of information and entertainment programming. Its customers enjoy many channels of news, sports, movies, comedy, drama, and educational programming with the simplicity and reliability of cable's broadband, fiber-optic network. Insight is also developing and providing the services that are becoming part of everyone's future—high-speed Internet access, digital cable, and telephony service.

Innovation in Infrastructure

When Insight took over the Louisville area's cable operations in late 1999, customers noticed a quick improvement in services, followed by an enhanced variety of options. The infrastructure of Insight's cable system allows customers to select from an extensive menu of services that are continuously updated as improved technology becomes available. By implementing this cutting-edge system, Insight has protected customers from hardware obsolescence. This system enables the company to deploy new technology with all of the latest advancements in entertainment and communication services. Furthermore, Insight's technology means that customers do not need a satellite dish or antenna for clear reception of local channels.

Insight offers channels with an array of entertainment and information options. Basic cable service provides clear reception of local broadcast and cable channels, and connects customers to additional services. Classic cable service offers the most popular cable networks available today, with a broad range of programming for every member of the family. Numerous premium movie channels are also available.

In addition, with Insight Home Theater, pay-per-view movies and events are offered to customers on several channels operating 24 hours a day.

In 1999, Insight's Louisville system rebuild was completed in a $150 million upgrade that increased bandwidth capacity to 750 megahertz. The resulting two-way fiber-optic system delivers a clearer digital picture with fewer disruptions and doubles the capacity of the former channel lineup, both analog and digital. The upgrade also introduced two revolutionary products into the Insight family of services. The first is LocalSource[SM], which is an interactive community information and entertainment guide customized exclusively for Louisville-area customers. They can use LocalSource to access local weather information, sports, cinema listings, restaurant menus, school activities, and much more. The second new product is a video-on-demand service called OnSet On Demand TV, which downloads current movies into the digital set-top box, so customers can watch any available movie at any time, with the ability to rewind, fast-forward, and pause.

This system upgrade also gives Insight customers the opportunity to enjoy high-speed Internet access

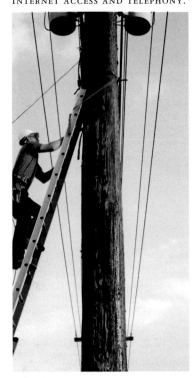

Clockwise from top: Customers of Insight Communications have come to expect a lot more from this company than basic cable service. One of the nation's largest cable operators, Insight offers its customers a variety of state-of-the-art entertainment, educational, and information services in each of the many communities it serves.

Local high school students film a segment about the C-SPAN school bus for their in-house news program. The bus is a mobile television production studio that tours the country and demonstrates the cable industry's commitment to education.

Cable has emerged as a flexible and efficient pipeline into consumers' homes for a full range of telecommunications services, including high-speed Internet access and telephony.

Louisville

through the same cable lines that provide their television service. Called Insight@Home, this service gives Insight customers a range of data services at speeds up to 100 times faster than traditional dial-up phone lines, reducing the time needed to download larger digital files.

Joint Ventures, Improved Services

In mid-1999, plans were under way for Insight to join forces with AT&T to provide telephony services to Insight customers. With Insight's expertise in cable television and AT&T's proven experience in telecommunications, the resources are solidly in place as the industry continues to move forward. Customers who choose this telephony option will receive all their cable television, Internet, and telephone services on the same cable line. Representative of the very latest in integrated technology, this project demonstrates Insight's commitment to providing the innovations that customers expect.

Yet all this technology does not mean very much if customers do not receive quick and courteous technical support when they need it. Therefore, Insight makes customer service its highest priority. Among its customer service commitments, Insight participates in the cable industry's On-Time Guarantee initiative, crediting customers' accounts if installers fail to arrive on time for a scheduled installation or service appointment. Insight also makes every effort to accommodate customer schedules for service and repair appointments, and is always committed to improving customer service in additional ways.

Insight Communications is an active member of the Louisville-area community. The company offers the use of its channels to local programs that provide heightened awareness and information about community events, issues, and activities. Through an industry-wide initiative called Cable in the Classroom, Insight also provides free cable service to every elementary, junior, and senior high school in the communities it serves, allowing teachers to create new curricula and innovative learning environments by accessing cable programming. In addition, Insight employees regularly participate in community activities such as Race for the Cure, collecting toys for Kosair Charities, and Derby Festival activities.

As Insight continues to grow, it upholds the values upon which its initial success was based. The company makes significant investments in its people, its operations, and its future to realize its vision: becoming a multifaceted communications business driven by quality customer service.

The Cartoon Network's traveling van attracts a crowd at the Louisville Zoo.

Insight's award-winning productions team edits local programming.

The Greatest City

Towery Publishing, Inc.

Beginning as a small publisher of local newspapers in the 1930s, Towery Publishing, Inc. today produces a wide range of community-oriented materials, including books (Urban Tapestry Series), business directories, magazines, and Internet publications. Building on its long heritage of excellence, the company has become global in scope, with cities from San Diego to Sydney represented by Towery products. In all its endeavors, this Memphis-based company strives to be synonymous with service, utility, and quality.

A Diversity of Community-Based Products

Over the years, Towery has become the largest producer of published materials for North American chambers of commerce. From membership directories that enhance business-to-business communication to visitor and relocation guides tailored to reflect the unique qualities of the communities they cover, the company's chamber-oriented materials offer comprehensive information on dozens of topics, including housing, education, leisure activities, health care, and local government.

In 1998, the company acquired Cincinnati-based Target Marketing, an established provider of detailed city street maps to more than 200 chambers of commerce throughout the United States and Canada. Now a division of Towery, Target offers full-color maps that include local landmarks and points of interest, such as recreational parks, shopping centers, golf courses, schools, industrial parks, city and county limits, subdivision names, public buildings, and even block numbers on most streets.

In 1990, Towery launched the Urban Tapestry Series, an award-winning collection of oversized, hardbound photojournals detailing the people, history, culture, environment, and commerce of various metropolitan areas. These coffee-table books highlight a community through three basic elements: an introductory essay by a noted local individual, an exquisite collection of four-color photographs, and profiles of the companies and organizations that animate the area's business life.

To date, more than 80 Urban Tapestry Series editions have been published in cities around the world, from New York to Vancouver to Sydney. Authors of the books' introductory essays include former U.S. President Gerald Ford (Grand Rapids), former Alberta Premier Peter Lougheed (Calgary), CBS anchor Dan Rather (Austin), ABC anchor Hugh Downs (Phoenix), best-selling mystery author Robert B. Parker (Boston), *American Movie Classics* host Nick Clooney (Cincinnati), Senator Richard Lugar (Indianapolis), and Challenger Center founder June Scobee Rodgers (Chattanooga).

To maintain hands-on quality in all of its periodicals and books, Towery has long used the latest production methods available. The company was the first production environment in the United States to combine desktop publishing with color separations and image scanning to produce finished film suitable for burning plates for four-color printing. Today, Towery relies on state-of-the-art digital prepress services to produce more than 8,000 pages each year, containing well over 30,000 high-quality color images.

An Internet Pioneer

By combining its long-standing expertise in community-oriented

Towery Publishing President and CEO J. Robert Towery has expanded the business his parents started in the 1930s to include a growing array of traditional and electronic published materials, as well as Internet and multimedia services, that are marketed locally, nationally, and internationally.

published materials with advanced production capabilities, a global sales force, and extensive data management capabilities, Towery has emerged as a significant provider of Internet-based city information. In keeping with its overall focus on community resources, the company's Internet efforts represent a natural step in the evolution of the business.

The primary product lines within the Internet division are the introCity™ sites. Towery's introCity sites introduce newcomers, visitors, and longtime residents to every facet of a particular community, while simultaneously placing the local chamber of commerce at the forefront of the city's Internet activity. The sites include newcomer information, calendars, photos, citywide business listings with everything from nightlife to shopping to family fun, and on-line maps pinpointing the exact location of businesses, schools, attractions, and much more.

Decades of Publishing Expertise

In 1972, current President and CEO J. Robert Towery succeeded his parents in managing the printing and publishing business they had founded nearly four decades earlier. Soon thereafter, he expanded the scope of the company's published materials to include *Memphis* magazine and other successful regional and national publications. In 1985, after selling its locally focused assets, Towery began the trajectory on which it continues today, creating community-oriented materials that are often produced in conjunction with chambers of commerce and other business organizations.

Despite the decades of change, Towery himself follows a long-standing family philosophy of unmatched service and unflinching quality. That approach extends throughout the entire organization to include more than 120 employees at the Memphis headquarters, another 80 located in Northern Kentucky outside Cincinnati, and more than 40 sales, marketing, and editorial staff traveling to and working in a growing list of client cities. All of its products, and more information about the company, are featured on the Internet at www.towery.com.

In summing up his company's steady growth, Towery restates the essential formula that has driven the business since its first pages were published: "The creative energies of our staff drive us toward innovation and invention. Our people make the highest possible demands on themselves, so I know that our future is secure if the ingredients for success remain a focus on service and quality."

TOWERY PUBLISHING WAS THE FIRST PRODUCTION ENVIRONMENT IN THE UNITED STATES TO COMBINE DESKTOP PUBLISHING WITH COLOR SEPARATIONS AND IMAGE SCANNING TO PRODUCE FINISHED FILM SUITABLE FOR BURNING PLATES FOR FOUR-COLOR PRINTING. TODAY, THE COMPANY'S STATE-OF-THE-ART NETWORK OF MACINTOSH AND WINDOWS WORKSTATIONS ALLOWS IT TO PRODUCE MORE THAN 8,000 PAGES EACH YEAR, CONTAINING MORE THAN 30,000 HIGH-QUALITY COLOR IMAGES (TOP).

THE TOWERY FAMILY'S PUBLISHING ROOTS CAN BE TRACED TO 1935, WHEN R.W. TOWERY (FAR LEFT) BEGAN PRODUCING A SERIES OF COMMUNITY HISTORIES IN TENNESSEE, MISSISSIPPI, AND TEXAS. THROUGHOUT THE COMPANY'S HISTORY, THE FOUNDING FAMILY HAS CONSISTENTLY EXHIBITED A COMMITMENT TO CLARITY, PRECISION, INNOVATION, AND VISION (BOTTOM).

LIBRARY OF CONGRESS CATALOGING-IN-PUBLICATION DATA
Louisville : the greatest city / [introduction] by Muhammad Ali ; art direction by Bob Kimball.
 p. cm. — (Urban Tapestry Series)
 "Sponsored by Greater Louisville Inc."
 Includes index.
 ISBN 1-881096-85-8 (alk. paper)
 1. Louisville (Ky.)—Civilization. 2. Louisville (Ky.)—Pictoral works. 3. Louisville (Ky.)—Economic conditions. 4. Business enterprises—Kentucky—Louisville. I. Ali, Muhammad, 1942- II. Greater Louisville Inc. III. Series.

F459.L85 L68 2000   00-062990
976.9'44—dc21

Printed in Spain

COPYRIGHT © 2000 BY TOWERY PUBLISHING, INC.
All rights reserved. No part of this work may be reproduced or copied in any form or by any means, except for brief excerpts in conjunction with book reviews, without prior written permission of the publisher.

TOWERY PUBLISHING, INC.
The Towery Building
1835 Union Avenue
Memphis, TN 38104
www.towery.com

PUBLISHER: J. Robert Towery EXECUTIVE PUBLISHER: Jenny McDowell NATIONAL SALES MANAGER: Stephen Hung MARKETING DIRECTOR: Carol Culpepper PROJECT DIRECTORS: Theresa Adkins, Candice Gilbert, Dawn Park-Donegan, Capi Porter EXECUTIVE EDITOR: David B. Dawson MANAGING EDITOR: Lynn Conlee SENIOR EDITORS: Carlisle Hacker, Brian L. Johnston EDITORS: Jay Adkins, Stephen M. Deusner, Rebecca E. Farabough, Sabrina Schroeder, Ginny Reeves EDITOR/CAPTION WRITER: Sunni Thompson COPY EDITOR: Danna M. Greenfield EDITORIAL ASSISTANT: Andrew S. Harlow PROFILE WRITER: Larissa Reece CREATIVE DIRECTOR: Brian Groppe PHOTOGRAPHY EDITOR: Jonathan Postal PHOTOGRAPHIC CONSULTANT: Dan Dry & Associates PROFILE DESIGNERS: Rebekah Barnhardt, Laurie Beck, Laura Higley, Glen Marshall PRODUCTION MANAGER: Brenda Pattat PHOTOGRAPHY COORDINATOR: Robin Lankford PRODUCTION ASSISTANTS: Robert Barnett, Loretta Lane DIGITAL COLOR SUPERVISOR: Darin Ipema DIGITAL COLOR TECHNICIANS: Eric Friedl, Brent Salazar COLOR SCANNING TECHNICIANS: Brad Long, Mark Svetz PRODUCTION RESOURCES MANAGER: Dave Dunlap Jr. PRINT COORDINATOR: Beverly Timmons

© WEASIE GAINES / DAN DRY & ASSOCIATES

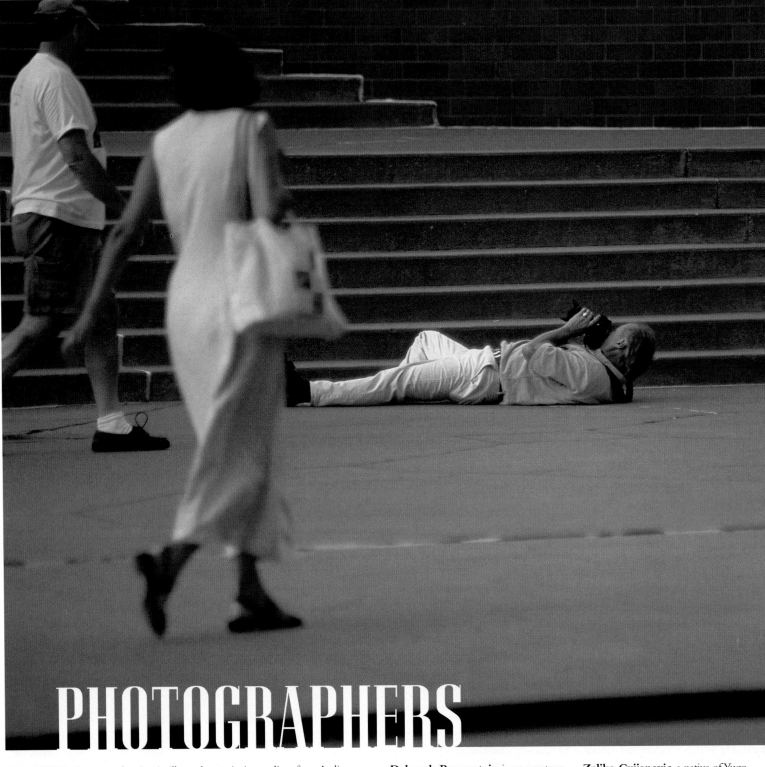

PHOTOGRAPHERS

Ronald W. Bailey moved to Louisville in 1969 after spending his childhood as a military brat. He attended Jefferson County Community College and the University of Louisville before focusing his attention on photography. Bailey now works for Photorific Images, where he specializes in nature, wildlife, and action photography.

Steve Baker is an international photographer who has contributed to more than 100 publications. With a degree in journalism from Indiana University, he is the proprietor of Highlight Photography, specializing in assignments for clients such as Eastman Kodak, Nike, Budweiser, the U.S. Olympic Committee, and Mobil Oil, which has commissioned seven exhibitions of his work since 1994. Baker is an Indianapolis resident and is the author and photographer of *Racing Is Everything*. He has contributed to numerous Towery Publications.

Deborah Brownstein is an amateur photographer who specializes in nature, floral, animal, and human interest photography. As president of the Louisville Photographic Society, Brownstein has been the winner of numerous local and state photography competitions.

Todd Bynum is a professional photographer residing in Jackson, Tennessee, where he specializes in portrait and equine photography.

Zeljko Cvijanovic, a native of Yugoslavia, now makes his home in Louisville. He is the owner and operator of Photography by Zeljko.

Linda Doane received a bachelor of science degree and a master of science degree from the University of Wisconsin before moving to Louisville in 1997. As the owner and operator of a private studio in Shelbyville, she specializes in portrait, wedding, and commercial photography. Doane's work has appeared in

which have now been published in more than 20 nations. Faris also completed an art project for the Hoosier Salon with a grant from the Indiana Arts Commission and the National Endowment for the Arts. Faris' images have appeared in numerous Towery publications.

Larry S. Foster moved to the Louisville area in 1973, and attended Jefferson County Community College and the University of Louisville. A two-time winner of the *Courier-Journal Scene* magazine photo contest, he specializes in nature and human interest photography.

Bob Helvey is a financial planner working for the Nowacki Agency. As an amateur photographer, he has won several ribbons at the Kentucky State Fair and has been published in *Kentucky Living*. Helvey specializes in people, landscape, flower, and candid photography.

Gertrude Hudson, originally from Washington County, Kentucky, moved to Louisville in 1952. Although she has no formal training in photography, Hudson loves to take pictures, especially nature shots. She also enjoys traveling, both in the United States and overseas.

Marshall Johnson, a lifelong resident of Louisville, concentrates his photography on landscapes and nature. He has had prints published in *The Voice-Tribune* and won first place in the Marie Shelby Botanical Gardens Photographic Exhibition in Sarasota, Florida. Johnson is also an active member of the Louisville Photographic Society.

Dorothy Johnston is a freelance writer and photographer for the *Cincinnati Post* and the *Kentucky Post*. She won the Kentucky Press award for best feature story in a class III weekly for 1996, and her photography can also be seen in *Northern Kentucky: Looking to the New Millennium*. Johnston specializes in event photography and feature story journalism.

Mary E. Krider is a native of Louisville and a 1968 graduate of Bellarmine-Ursuline College. As a photo journalist and nature photographer, she has worked for *Today's Woman*, *The Voice-Tribune*, *Jeffersontown News Leader*, and *Fern Creek News Leader*. Krider photographed the 1992 Olympic Games in Barcelona and the 1996 Games in Atlanta.

Linda Morton is a freelance photographer specializing in human interest, event, and photo-essay photography. She has worked for Iroquois Amphitheatre, Presbyterian Community Center, and private clients in and around Louisville. Morton has had work featured in *The Voice-Tribune*, *Courier-Journal*, and *Kentucky Living*.

Judi Parks is an award-winning photojournalist and professional writer living and working in the San Francisco Bay Area. Her work has been collected by numerous museums and public collections in the United States and Europe, and can also be seen in Towery Publishing's *Fresno: Heartbeat of the Valley*. Her documentary series *Home Sweet Home: Caring for America's Elderly* was honored with the *Communication Arts-Design Annual* 1999 Award of Excellence for an unpublished series.

Photophile, established in San Diego in 1967, has more than 1 million color images on file, culled from more than 85 contributing local and international photographers. Subjects range from images of Southern California to adventure sports, wildlife and underwater scenes, business, industry, people, science and research, health and medicine, and travel photography. Included on Photophile's client list are American Express, *Guest Informant*, Franklin Stoorza, and Princess Cruises.

Ron L. Profumo earned a bachelor of science degree in elementary education from the University of Louisville before completing a master's degree in child guidance at Western Kentucky University. He is a full-time employee of Jefferson County Public Schools. In his free time, Profumo focuses his photography on horse racing and human interest subjects.

William L. Rhodes, a retired naval engineer, now spends his time photographing outdoor events, fairs, carnivals, and parades. He also photographs historical relics such as trains, tractors, and steam-powered boats.

Lynn T. Shea, a native of Frankfort, now makes her home in the Louisville area. As a self-taught photographer, Shea has won many awards, both nationally and internationally, and has been featured in several publications.

Donald J. Sivori specializes in landscape and nature photography. He is a past winner of the *Courier-Journal Scene* magazine photo contest and a member of the Louisville Photo Forum Camera Club.

Jeffrey L. Vaughn completed a bachelor of fine arts degree at Washington University in St. Louis before moving to the University of Dallas, where he earned a master of arts degree and a master of fine arts degree. With recent exhibits at Locus Gallery in St. Louis and Steinway Gallery in Chapel Hill, Vaughn has featured his work in national galleries regularly since 1979.

Other photographers who have contributed to *Louisville: The Greatest City* include **Ted Bressoud**, **Michael Clevenger**, **Jeffrey Evans**, **Sandy Nabb**, and **Tsung-Yao Huang**. Please contact Towery Publishing for additional information on photographers with images appearing in this publication.

magazines including *Backstretch* and *Bloodhorse*.

Charlene Faris, a native of Fleming County, Kentucky, is the owner and operator of Charlene Faris Photos. Specializing in travel, historic, and inspirational photography, Faris has won numerous awards, including several from the National League of American Pen Women art shows. She was a 1994 Pulitzer Prize nominee for wedding photos of Lyle Lovett and Julia Roberts,

Index of Profiles

AAF International	318
Adelphia Business Solutions	414
ADVENT	406
Aegon Insurance Group	308
Anthem Blue Cross and Blue Shield in Kentucky	322
Bakery Chef, Inc.	374
Bellarmine University	324
Brown, Todd & Heyburn PLLC	357
Business First	382
Cambridge Construction Company	408
Carlson Wagonlit Travel/WTS	354
Cintas Corporation	394
Classroom Teachers Federal Credit Union	332
Clear Channel Communications, Inc.	362
Dismas Charities, Inc.	346
Fire King International Inc.	326
Ford Motor Company	312
GE Appliances	325
Goldberg & Simpson, PSC	376
Gordon Insurance Group	364
Greater Louisville Inc.	416
Heick, Hester & Associates	384
Humana	340
Infinity Outdoor	342
Insight Communications	422
Jewish Hospital HealthCare Services	304
Johnson Controls	398
Kentucky Fair and Exposition Center/Kentucky International Convention Center	302
The Kroger Company	296
L&N Federal Credit Union	334
LG&E Energy	288
LabCorp	344
Lear Corporation	396
Lightyear	412
Louisville and Jefferson County Convention & Visitors Bureau	349
Louisville/Jefferson County Metropolitan Sewer District	315
Manpower Inc.	320
Micro Computer Solutions	380
Muhammad Ali Center	420
Musselman Hotels	400
NTS Development Company	350
Neace Lukens	402
Norton Healthcare	298
OPM Services, Inc.	410
Papa John's International	386
Park Federal Credit Union	352
Peter Built Homes, Inc.	316
Presbyterian Church (USA)	392
PRIMCO Capital Management	388
Prudential Parks & Weisberg Realtors®	328
Roman Catholic Archdiocese of Louisville	284
St. Xavier High School	290
Samtec, Inc.	366
Steel Technologies Inc.	358
Süd-Chemie Inc.	368
The Sullivan Colleges System	292
Sun Properties	393
SYSCO	356
TechRepublic	418
Thomas Industries Inc.	336
Torbitt & Castleman	294
Towery Publishing, Inc.	424
Trinity High School	330
United Parcel Service (UPS)	378
Vincenzo's	390
WDRB Fox 41	360
Wyatt, Tarrant & Combs	286